THE BRAIN AND REWARD

THE BRAIN AND REWARD

BY

E. T. ROLLS, M.A. (Cantab.); D.Phil. (Oxon.)

University of Oxford, Department of Experimental Psychology

PERGAMON PRESS

OXFORD · NEW YORK · TORONTO · SYDNEY · BRAUNSCHWEIG

Pergamon Press Ltd., Headington Hill Hall, Oxford

Pergamon Press Inc., Maxwell House, Fairview Park, Elmsford, New York 10523

Pergamon of Canada Ltd., 207 Queen's Quay West, Toronto 1

Pergamon Press (Aust.) Pty. Ltd., 19a Boundary Street, Rushcutters Bay, N.S.W. 2011, Australia

Pergamon Press GmbH, Burgplatz 1, Braunschweig 3300, West Germany

---

First edition 1975

Library of Congress Cataloging in Publication Data

Rolls, E T

The brain and reward.

Bibliography: p.

1. Brain stimulation. 2. Reward (Psychology)
I. Title. [DNLM: 1. Electrophysiology. 2. Neurophysiology. 3. Reward.
WL102 R755b]
QP388.R64 1975 612'.82 74-32290
ISBN 0-08-018225-9 pbk.

Printed in Great Britain by A. Wheaton & Co., Exeter

CONTENTS

ACKNOWLEDGEMENT

THE author has worked with M. J. Burton, S. J. Cooper, A. Edelson, C. R. Gallistel, B. P. Jones, P. H. Kelly, B. J. Rolls, M. Rush, and S. G. Shaw, and their collaboration in many of the experiments described here is sincerely acknowledged.

CHAPTER 1

INTRODUCTION

1.1. GENERAL INTRODUCTION

Electrical stimulation of certain regions of the brain is rewarding in that animals and man will learn a task to obtain the stimulation in much the same way as they will learn a task to obtain food. Terms used to describe this are intracranial self-stimulation and brain-stimulation reward.

Brain-stimulation reward was discovered by Olds and Milner (1954) during an investigation of the effects of electrical stimulation of the septal area of the brain in rats. A rat would return persistently to a place where the stimulation had been given previously. The attractiveness of the stimulation for the rat was demonstrated more objectively by connecting a Skinner box so that the rat delivered stimulation to itself whenever it pressed the bar (Fig. 1). Rats with stimulation electrodes in certain parts of the brain bar-pressed very vigorously and persistently to obtain the stimulation (Fig. 2).

One area of the brain in which good self-stimulation is found is the lateral hypothalamus. This area is involved in the control of feeding in that bilateral lesions of the lateral hypothalamus produce a prolonged decrease of food intake (Fig. 3). Lateral hypothalamic self-stimulation is closely related to feeding in that hunger increases lateral hypothalamic self-stimulation rate. Hunger also increases the rate of bar-pressing for food. Thus lateral hypothalamic stimulation appears to mimic food given to an animal. The way in which lateral hypothalamic stimulation is equivalent to a food reward, and its use in analysing eating, drinking, and other types of motivated behaviour, are described in Chapter 2. For example, it is shown that sensory inputs such as taste, smell, and vision normally guide animals to food

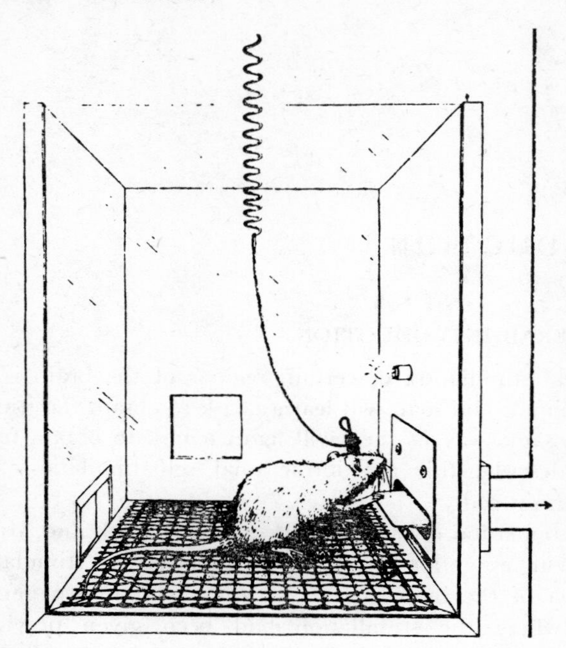

FIG. 1. The Skinner box is connected so that every time the rat presses the bar a short train (e.g. 0.5 s) of pulses of electrical stimulation is delivered to an area of the brain through an implanted electrode. Brain-stimulation reward is said to occur, and can be measured, if the rat learns to bar-press and bar-presses repeatedly to obtain the stimulation. (Reproduced with permission from R. F. Thompson, *Foundations of Physiological Psychology*, Harper & Row, 1967.)

or water, and that food and water intake are normally only maintained if the sensory inputs are present. In other words, the sensory inputs provide reward or reinforcement. The sensory inputs produce firing in hypothalamic cells. In recent experiments in my laboratory it has been found that these cells are also fired by brain-stimulation reward. Thus an explanation of brain-stimulation reward is that it activates cells which normally signal natural reward. The hypothalamic cells are only strongly activated by the taste or smell of food if the animal is hungry. Thus the factors which indicate hunger (such as the glucose level in the blood or body weight) must gate or control the effect which sensory input has on these cells. This gating of

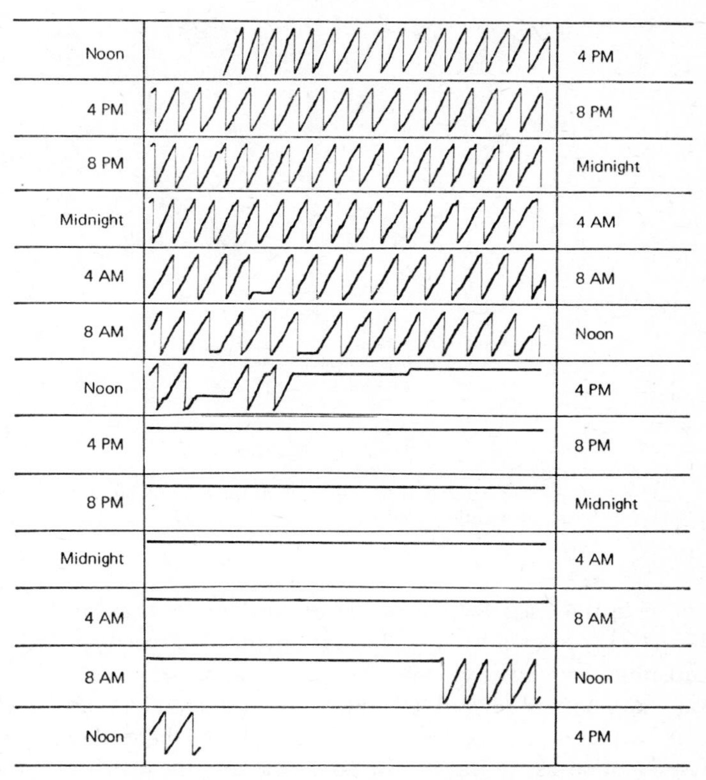

FIG. 2. This cumulative record shows that intracranial self-stimulation can be very vigorous and persistent. Each time the rat bar-pressed (for a 0.5 s train of electrical stimulation) the pen stepped once vertically. After 500 steps it resets to 0. (From J. Olds, Satiation effects in self-stimulation of the brain, *J. Comp. Physiol. Psychol.* **51,** 675–8, 1958. Copyright 1958 by the American Psychological Association, and reproduced by permission.)

sensory input by hunger probably makes lateral hypothalamic stimulation rewarding only in a hungry animal, and lateral hypothalamic stimulation (and food) aversive in an overfed animal. This type of gating of the effect of sensory input on hypothalamic neurones may

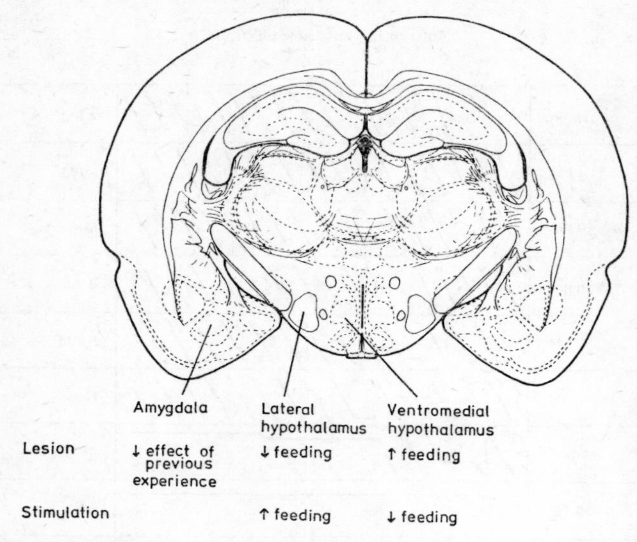

FIG. 3. Coronal (transverse) section of the rat brain at the hypothalamic level showing how lesions and stimulation affect feeding.

also provide the basis for the observations that the taste and smell of food are very pleasant in a hungry man, and become gradually neutral and then aversive when hunger is gradually reduced.

Brain-stimulation reward can also produce emotions and feelings such as pleasure and happiness in man, and can therefore be used to analyse emotion as well as motivation. One area in which stimulation can produce this type of general emotion is in the temporal lobe, in or near the amygdala. The role of the amygdala (as well as other parts of the brain) in brain-stimulation reward is considered in Chapter 3. Many observations in animals show that the amygdala is involved in self-stimulation. For example, amygdaloid neurones are activated in self-stimulation, anaesthetization or lesions of the amygdala disrupt self-stimulation at least temporarily, and self-stimulation of the amygdala occurs. The amygdala may be involved in brain-stimulation reward because it is normally involved in learning about rewards (and punishments). For example, animals with bilateral damage to the amygdala learn only poorly which visual stimulus is associated with food or which auditory stimulus is associated with electric shock. Thus the lesioned animals have difficulty in recognizing which are food objects, and repeatedly place non-food as well as food objects in their mouths. This type of learning is clearly important in food and water intake, and has only recently received attention. The nature of the learning impairment of animals with amygdala lesions also means that they are tame—they do not associate the sight of a frightening stimulus with fear. Thus research on the amygdala suggests that it is crucial for much emotional behaviour in that it is involved in learning which stimuli are rewarding (or pleasant) and which stimuli are aversive (or unpleasant). In view of these hypotheses it is of interest that neurones in the amygdala appear to learn which environmental stimuli are rewarding, and which are aversive, so that the amygdala can be considered as a region which receives sensory inputs and which connects these to reward or punishment if appropriate (see Chapter 3). It may be that because of amygdalo-hypothalamic connections some hypothalamic food-reward neurones can be fired by the sight of food, and that environmental stimuli can produce pleasant or unpleasant emotional feelings.

The orbitofrontal cortex is also involved in brain-stimulation reward (see Chapter 3). Its relation to brain-stimulation reward may be that it is involved in altering behaviour when stimuli are no longer followed by reward (or punishment). Thus this region also appears to be crucial in emotional behaviour—malfunction can lead to persevering responsiveness to stimuli which previously produced emotions (see Chapter 3).

As brain-stimulation reward is closely related to emotion, brain-stimulation reward has also been used to analyse the pharmacology of emotion. It is clear that a disturbed reward (or punishment) system will lead to abnormal emotional behaviour, and extrapolations of this research to emotional disorders such as depression have already been made. But much more work in this rapidly expanding research area is necessary before any firm conclusions can be made.

Phenomena which have been considered to be peculiar to brain-stimulation reward are discussed in Chapter 5. First, in Chapter 1, the sense in which electrical stimulation of the brain provides reward and the way in which the electrical stimulation affects the brain are discussed.

Brain-stimulation reward has been found in many vertebrates. Examples (for references see section 3.2) are the goldfish, pigeon, rabbit, cat, dog, dolphin, monkey (Routtenberg *et al.*, 1971; Rolls, 1974), and man (Heath, 1954, 1963, 1964; Higgins *et al.*, 1956; Delgado, 1960; Delgado and Hamlin, 1960; Heath and Mickle, 1960; Sem-Jacobsen and Torkildsen, 1960; Bishop *et al.*, 1963; Sem-Jacobsen, 1968; Delgado, 1969; Mark *et al.*, 1972). Most of the observations on man were made during the course of neurosurgical investigation or treatment of tumours, epilepsy, schizophrenia, or Parkinson's disease. In these studies it was possible not only to demonstrate intracranial self-stimulation, but also to ask the patient about the effects of the stimulation. In many sites the stimulation was said to feel good, to produce pleasure, and to produce relaxation. The regions where stimulation has been found in the above species to produce reward are typically deep in the brain, in or near the hypothalamus, and in limbic and related structures, e.g. the amygdala deep in the temporal lobe, and the orbitofrontal cortex (see Chapter 3).

1.2. EVIDENCE THAT REWARD IS MEDIATED BY THE ELECTRICAL STIMULATION

The electrical stimulation of the brain can provide reward in that animals will learn a task and work to obtain the stimulation. The animals are clearly not caught up in a motor loop in which the stimulation forces them to make the movement that results in another bar-press. This is shown in the original discovery of self-stimulation, in which rats returned to a place in an open field where stimulation had been given previously (Olds and Milner, 1954; Milner, 1970). Similarly, rats will run through a maze in order to obtain the stimulation (Olds, 1956). The behaviour of the rats in the maze is similar to that of hungry rats running for a reward of food at the end of the maze (Fig. 4). Rats will also cross an electrified grid which produces foot-shock to obtain electrical stimulation of reward sites (Fig. 5; Olds, 1956, 1961). These experiments, together with the finding that man will work for (Bishop *et al.*, 1963; Heath, 1963) or ask for

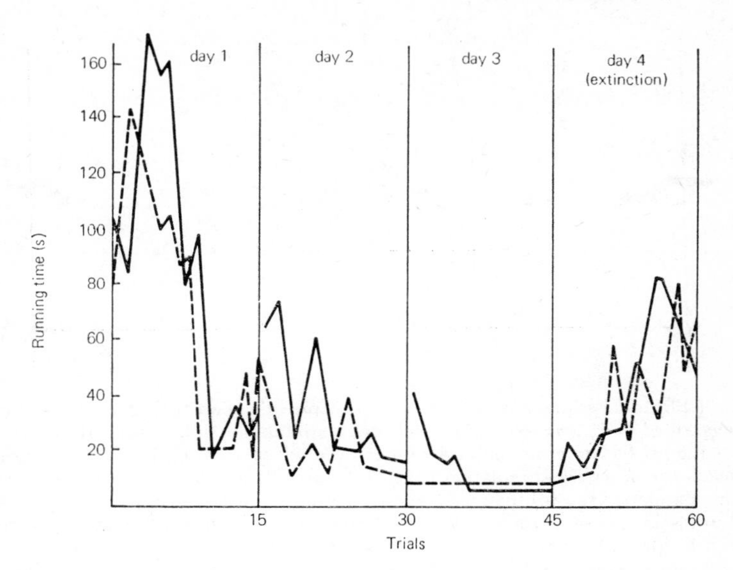

FIG. 4. Comparability of learning to run a maze for brain-stimulation reward (——, 8 rats) and for food reward (– – – –, 7 rats). The running times decreased similarly for brain-stimulation and food reward on days 1–3 and increased similarly when brain-stimulation and food reward were omitted on day 4. (From J. Olds, Runway and maze behavior controlled by basomedial forebrain stimulation in the rat, *J. Comp. Physiol. Psychol.* **49,** 507–12, 1956. Copyright 1956 by the American Psychological Association, and reproduced by permission.)

stimulation of some brain areas (Delgado, 1969) and finds the stimulation pleasurable (see references in sections 1.1 and 2.2), show that electrical stimulation of the brain can provide reward.

The effectiveness of the electrical stimulation can be measured in a number of ways. The usual way is to arrange for a short train (e.g. 0.3 or 0.5 s long) of brief pulses at 100 pulses/s to be given for every bar-press. To obtain the stimulation the animal has to bar-press, and the rate of bar-pressing gives a measure of the effectiveness of the stimulation. A similar measure is the lowest stimulation current which will maintain a given rate of self-stimulation. These measures are likely to be affected by a number of factors, e.g. the degree of arousal or drowsiness of the animal. A different type of measure of the reward value of the stimulation is preference. In an experiment by

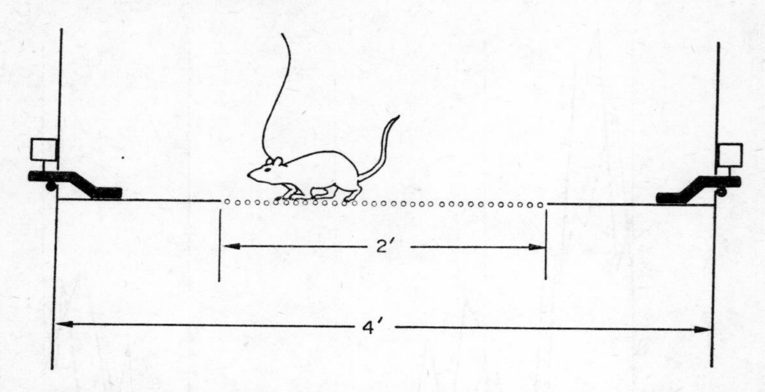

Fig. 5. Electrical stimulation of the brain can provide reward in that rats will cross an electrified shock grid to obtain the electrical stimulation of the brain. In this experiment the rat had to cross the grid to reach the self-stimulation lever, then recross it to reach the other self-stimulation lever. The rats tolerated more foot-shock to reach brain-stimulation reward than food reward. (From J. Olds, Differential effects of drive and drugs on self-stimulation at different brain sites, in *Electrical Stimulation of the Brain*, ed. D. E. Sheer, University of Texas Press, 1961.)

Hodos and Valenstein (1962) it was shown that rats could be made to choose stimulation of the septal area rather than stimulation of the lateral hypothalamus, yet would bar-press more rapidly for the lateral hypothalamic stimulation. Preference may be the better measure of reward as it controls for general factors such as arousal. Another measure of reward strength is the number of pulses of brain stimulation which are sufficient to maintain running, when the pulses are given as a reward at the end of a runway (Gallistel, 1969a, 1973). (This procedure is effective if care is taken to minimize intertrial interference by giving priming: see Gallistel, 1973, p. 204.) An example of this type of measurement of reward is shown in Fig. 6. As the number of stimulation pulses given per run is increased, running speed reaches a maximum. The number of pulses at which this occurs gives a measure of the reward value of the stimulation. Although there are many measures of the potency of rewarding brain stimulation, each measure may be affected by a number of factors, and care must always be taken when conclusions about reward (in contrast to self-stimulation rate, for example) are made.

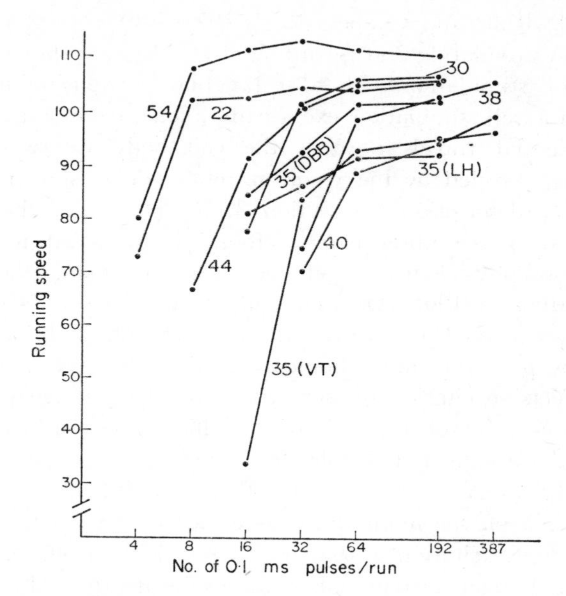

FIG. 6. Running speed along a runway as a function of the number of pulses given as a reward at the end of the runway for running. Rats 54 and 22 with lateral hypothalamic stimulation showed an asymptote of running speed with fewer reward pulses than were needed in rats 35 (VT) and 40 with ventral tegmental stimulation. This is one method by which the relative amounts of reward produced by stimulation at different sites can be compared. (From C. R. Gallistel, The incentive of brain-stimulation reward, *J. Comp. Physiol. Psychol.* **69,** 713–21, 1969. Copyright 1969 by the American Psychological Association, and reproduced by permission.)

1.3. NATURE OF THE ELECTRICAL STIMULATION WHICH PROVIDES REWARD

Electrical stimulation of the brain is applied through electrodes which are insulated except at the tip. The stimulation current passes through the tip, where it produces localized excitation of the brain. The most important factor is where the tip is located in the brain, and this is described in Chapter 3. The nature of the electrical stimulation is also important. Neurones are excited by the stimulation. This has been shown by recording from single neurones during stimulation of reward sites and during self-stimulation (Rolls, 1971a, c,

1972, 1974; Rolls and Cooper, 1973, 1974b). One possible arrangement of the stimulating and recording electrodes is as shown in Fig. 7. In this case stimulation with a rectangular current pulse near the threshold for self-stimulation excites the axon so that an all-or-none action potential travels towards the cell body where the action potential is recorded by the recording electrode as shown in Fig. 8. (For further description see section 3.7.3.) Details of the electrical properties of nerve fibres can be found in standard textbooks of physiology or physiological psychology, e.g. Grossman, 1973; Milner, 1970; Thompson, 1967; Deutsch and Deutsch, 1973. Neurones are most easily excited by negative pulses of current. Biphasic (negative followed by positive) pulses should be avoided because of the risk of anodal effects (excitation or depression) by the positive part of the pulse and because of the risk of metallic deposition with bipolar stimulation through polarizable (e.g. stainless-steel) electrodes (see Gallistel, 1973). A short pulse duration (e.g. 0.1 ms) is preferable because this is on the minimum charge part of the strength–duration curve of nerve excitation so that the current is most efficient at exciting nerves. (Longer current pulses also excite neurones, but relatively more energy is present and could produce tissue destruction by electrolysis or even heat production.) Short pulses also allow minimal accommodation, so that the complicating factor of anodal excitation

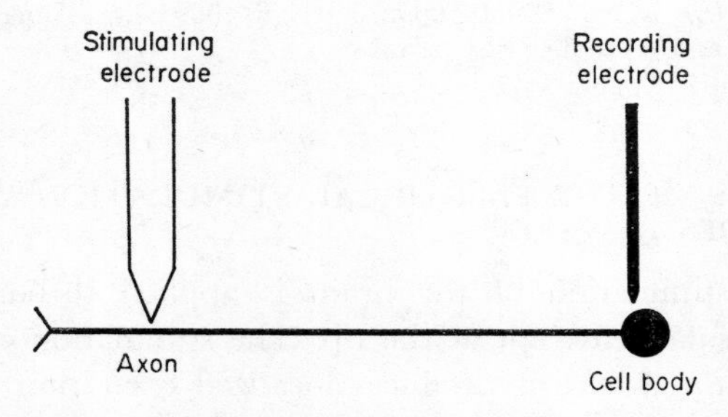

Fig. 7. Action potentials are normally generated at the cell body and travel orthodromically along the axon. If the axon is directly stimulated by an electrical pulse, then an action potential will travel antidromically towards the cell body and collide with and prevent the propagation of an action potential travelling normally away from the cell body. Whether collision occurs can be used to determine whether excitation is antidromic.

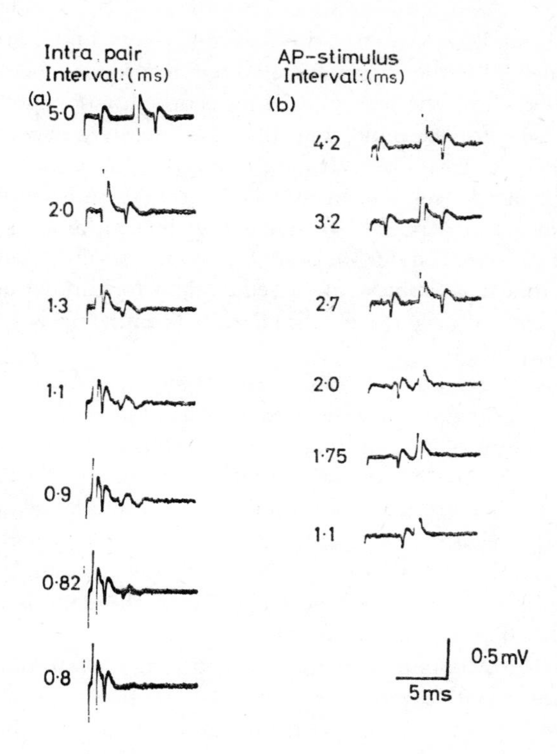

FIG. 8. Recording from a single unit in the sulcal prefrontal cortex directly excited by rewarding stimulation of the pons. (a) Stimulation with two pulses. *Top trace.* Each stimulus pulse is followed by an action potential (downward deflection) at a short, fixed latency of 2 ms. *Lower traces.* As the intrapair interval is reduced the second action potential is present until 0.82 ms, at which intrapair interval the second action potential occurs intermittently. At an intrapair interval of 0.8 ms an action potential never follows the second pulse, i.e. complete refractoriness is shown. (b) Collision evidence showing antidromic activation of the same neurone. When a spontaneous action potential preceded the twice-threshold stimulus pulse by 2.0 ms or less (AP-stimulus intervals of 2.0, 1.75, and 1.1 ms), the electrically elicited action potential (shown following the stimulus pulse in the upper three traces) failed to appear. Because the neurone had an absolute refractory period of 0.82 ms and was driven with a latency of 0.2 ms, collision of the electrically elicited action potential with the spontaneous action potential must have occurred. Therefore activation is antidromic. (From E. T. Rolls and S. J. Cooper, Connection between prefrontal cortex and pontine brain-stimulation reward sites in the rat, *Expl. Neurol.* **42,** 687–99, 1974.)

(produced by the positive-going part of the pulse) is avoided (see Hill, 1936; Gallistel, 1973). Short stimulus pulses also allow action potentials produced by the stimulation to be recorded clear of stimulus artefact, and allow the latency of the recorded action potential to be measured (see, for example, Fig. 8). It is usually necessary to excite the neurones at least several times, and 0.3 s trains of 0.1 ms pulses occurring every 10 ms (i.e. at 100 Hz) are very satisfactory. (Such a train contains thirty pulses, so that every time an animal bar-presses, it receives thirty stimulus pulses.) Sine-wave stimulation is unsatisfactory in that it is biphasic and is equivalent to stimulation with long pulses. A further account of stimulation parameters is contained in Gallistel (1973).

Monopolar electrodes, in which stimulation passes through the tip to the surrounding brain and thence to a large indifferent electrode located remotely on the animal, are preferable to bipolar electrodes. With bipolar electrodes, anodal excitation and other effects may occur at one of the electrodes in addition to the cathodal excitation produced by the negative pulses at the other electrode. This complicates the interpretation of the results in that, for example, the site of stimulation is uncertain, and electrolysis with the deposition of harmful metallic deposits from a polarizable anode can occur (Wetzel *et al.*, 1969). Monopolar stimulation through platinum electrodes with 0.1 ms negative pulses is thus very satisfactory, and even with stainless-steel electrodes no harmful deposits of metallic ions are produced, and polarization is insignificant as shown by the observation that the current threshold for the excitation of a neurone remains constant even when high intensity (2 mA), high frequency (100 Hz) stimulation is prolonged for several seconds (personal observation).

The distance around a stimulating electrode within which neurones are excited during self-stimulation is small (probably not more than 0.5 mm), but is not known exactly. The potential gradient on which excitation depends is greatest in the region of high current density close to the electrode tip, and falls off rapidly with increasing distance from the tip. A small electrode with a high impedance will allow the necessary potential gradient to be reached with a relatively small stimulation current, and the number of excited neurones for a given stimulating voltage is relatively small. In this way relatively localized stimulation can be used. (With self-stimulation it is usual for the voltage required to support self-stimulation to be relatively constant when different sizes, and thus impedances, of electrodes are used and other factors are held constant—personal observation.)

Using small electrodes, 62 μm wire with only the cross-section of the wire uninsulated, Olds *et al.* (1971) found drinking, feeding, and reward were elicited at different hypothalamic sites within 1 mm of each other. As mixed effects were not found, the stimulation must have been spreading less than 1 mm. Similarly, stimulation in the rat lateral hypothalamus elicits reward, yet if the current spread 1 mm medial to the ventromedial hypothalamus it would elicit aversion, and if it spread 1 mm lateral to the internal capsule it would elicit turning. Also, neurones recorded as close as 1 mm to self-stimulation electrodes are not directly excited by the rewarding stimulation (personal observation). These types of observation indicate that rewarding stimulation produces excitation in neurones only within a small distance of the tip. Of course, the excitation is then conducted by nerve impulses away from the stimulating site, along axons, and across synapses.

As shown in Fig. 8, neurones are excited during self-stimulation. Thus the electrical stimulation can mimic the increased firing of neurones which occurs during some types of behaviour (see, for example, Fig. 12) and can be used to analyse brain function. In that many neurones fire simultaneously with every current pulse, the pattern of firing is unlike that which occurs naturally. Thus although some disruption of normal activity must occur, it is also possible to mimic to some extent the effect of normal activity in a particular brain region. This is shown by the direct demonstration of the similarity of the effects of brain-stimulation and natural rewards on some neurones (see, for example, Fig. 12), and by the observations that lesions or anaesthetization of a region of the brain often have the opposite effect to electrical stimulation. For example, lesions or anaesthetization of the lateral hypothalamus stop eating, while electrical stimulation here elicits eating. Conversely, lesions or anaesthetization of the ventromedial hypothalamus produce overeating, while electrical stimulation here stops eating (see Fig. 3) (see, for example,

Epstein, 1960; Hoebel, 1969; Grossman, 1973, ch. 9). Thus electrical stimulation of the brain can mimic natural stimulation (see also Chapter 2), and often produces the opposite of the disruption of brain activity which follows a lesion.

1.4. SUMMARY

Electrical stimulation of the brain can provide reward in that animals will learn a task and work to obtain the stimulation. Comparable stimulation in man can produce emotional feelings such as pleasure. Trains (e.g. 0.3 s long) of short (e.g. 0.1 ms) negative pulses recurring at a high frequency (e.g. 100 pulses/s) are suitable for the stimulation. The stimulation excites neurones close to the stimulation electrode tip.

CHAPTER 2

THE NATURE OF THE REWARD PRODUCED

IN THIS chapter experiments which analyse the nature of the reward produced by electrical stimulation of reward sites in the brain are described. The first set of experiments shows that at some sites the stimulation is like a food reward given to a hungry animal—the stimulation can mimic a natural reward. The second set of experiments (section 2.1.2) shows that the stimulation can have the properties of a specific natural reward, e.g. at one site stimulation may mimic food, and at another site stimulation may mimic water.

In section 2.1.3 it is shown that brain-stimulation reward can excite neural pathways concerned with natural rewards by recording during self-stimulation from single cells in the hypothalamus which signal that food is available. Observations in man allow us to decide whether brain-stimulation reward can also mimic more general emotional states such as pleasure or happiness (section 2.2).

2.1. EVIDENCE FROM ANIMAL STUDIES

2.1.1. *Brain-stimulation reward can be equivalent to a natural reward*

If an animal is made hungry, then it will work harder to obtain food. Similarly, a hungry animal will work harder to obtain electrical stimulation of the lateral hypothalamus than will a satiated animal (Hoebel, 1969). In this respect, stimulation of some brain sites is equivalent to a natural food reward. Experiments which show this have been described by Hoebel (1969; see below).

In one experiment the injection of insulin (0.025 unit), which lowers blood sugar and causes hyperphagia, consistently increased lateral hypothalamic self-stimulation rate (Balagura and Hoebel, 1967; Fig. 9). Glucagon (0.25 mg), which increases blood sugar, decreased lateral hypothalamic self-stimulation (Fig. 9). Comparable results are that stomach distension decreased lateral hypothalamic reward (increased the current required to maintain self-stimulation; Hoebel, 1968), as did injections of liquid diet or glucose into the stomach (Mount and Hoebel, 1967; see also Fig. 10, in which self-stimulation rate was measured). If an animal is force-fed so that it becomes overweight, then it eats less food. It also self-stimulates less (MacNeil, 1966). When force-feeding is discontinued, body weight gradually decreases back to normal, and lateral hypothalamic self-stimulation rate gradually increases. Thus body weight, which appears

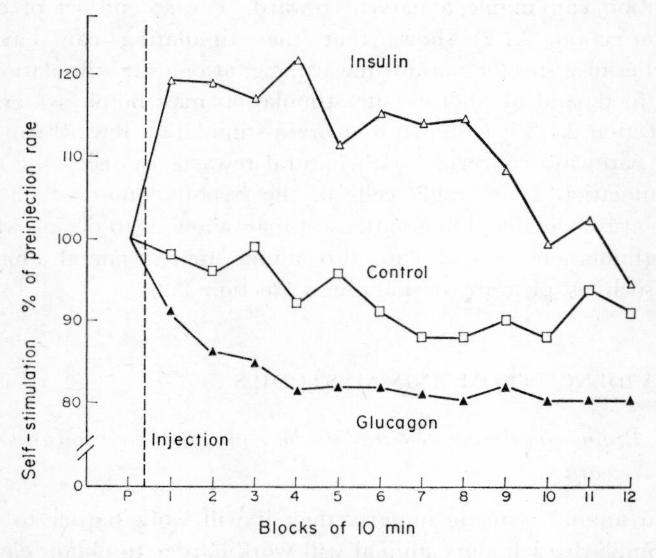

FIG. 9. Modulation of the reward value of lateral hypothalamic stimulation by a hunger signal. Subcutaneous injection of insulin (0.025 unit) increases lateral hypothalamic self-stimulation above the preinjection (P) level (mean values for 3 rats). Glucagon injections (0.25 mg) have the opposite effect (6 rats). The response to saline (open squares) is included as a control. (From S. Balagura and B. G. Hoebel, Self-stimulation of the hypothalamic "feeding-reward system" modified by insulin and glucagon, *Physiol. Behav.* **2,** 337–40, 1967.)

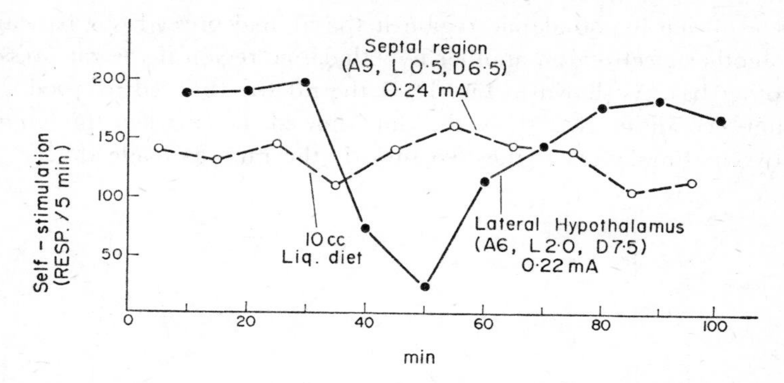

FIG. 10. Modulation of the reward value of stimulation occurs specifically at some sites. Differential inhibition of self-stimulation by tube-feeding liquid diet directly into the stomach of a rat. Self-stimulation slowed down at the lateral hypothalamic site but not at the septal site. The self-stimulation in this rat was measured at the two self-stimulation sites in alternate 5 min periods. (From B. G. Hoebel, Inhibition and disinhibition of self-stimulation and feeding, *J. Comp. Physiol. Psychol.* **66,** 89–100, 1968. Copyright 1968 by the American Psychological Association, and reproduced by permission.)

to control how rewarding food is, also controls the reward value of lateral hypothalamic stimulation (Hoebel, 1969).

In all these experiments the reward value of lateral hypothalamic stimulation appears to mimic the reward value of food. This indicates that lateral hypothalamic stimulation can be equivalent to a food reward.

2.1.2. *Brain-stimulation reward can be equivalent to a specific natural reward*

The next question is whether electrical stimulation of the brain can be equivalent to different types of natural reward, e.g. at some sites to food for a hungry animal, and at other sites to water for a thirsty animal. Evidence that brain-stimulation reward can be this specific was provided in an experiment by Gallistel and Beagley (1971). They allowed rats to choose between stimulation at two different hypothalamic sites. Stimulation was delivered to one electrode in one hypothalamic region if the animal pressed one bar, and to another electrode in another hypothalamic region if the rat pressed another bar. As shown in Fig. 11, if the rat was satiated for food and water (condition S), then the rat showed no marked preference between stimulation at the two sites. If the rat was made thirsty (T)

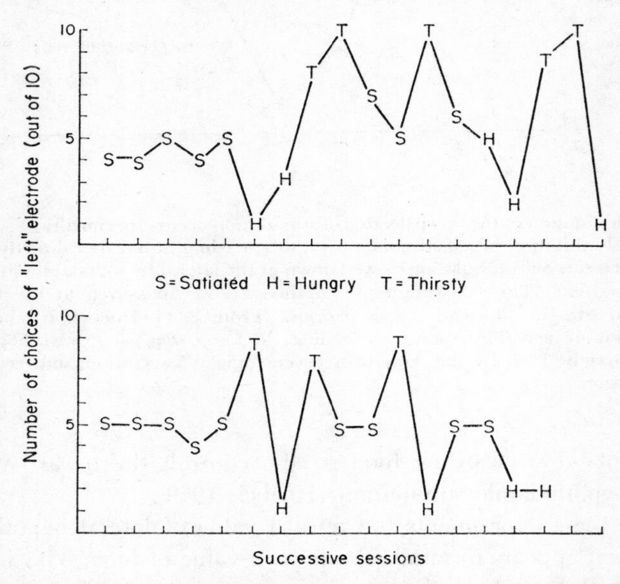

Fig. 11. Water and food reward effects produced by electrical stimulation of the brain in two rats. If the rats were thirsty (T) they chose to press a lever which delivered stimulation to one brain site, and if they were hungry (H) they chose to press the other lever which delivered stimulation to another brain site. S = satiated—neither hungry nor thirsty. (From C. R. Gallistel and G. Beagley, Specificity of brain-stimulation reward in the rat, *J. Comp. Physiol. Psychol.* **76,** 199–205, 1971. Copyright by the American Psychological Association, and reproduced by permission.)

by deprivation of water for 24 h, then it repeatedly chose stimulation at one hypothalamic site. If the rat was made hungry (H) by deprivation of food for 24 h, then it repeatedly chose stimulation at the other hypothalamic site. Thus the stimulation was equivalent to water at one site, and food at another site. Gallistel and Beagley (1971) were able to repeat this type of experiment in a number of different rats.

Another way of showing that brain stimulation can be equivalent to a specific natural reward is to measure self-stimulation rate at different sites after the appropriate type of deprivation. Although lateral hypothalamic self-stimulation rate is increased by hunger (Hoebel, 1969), it is not easy to show that the self-stimulation rate at any brain site is increased by thirst (Mogenson, 1969). But it has been possible to separate food-reward sites from sites where reward is modulated by sex hormones using self-stimulation rate. For example, Caggiula (1967, 1970) found in male rats that posterior hypothalamic self-stimulation rate was decreased by castration, and lateral hypothalamic self-stimulation rate was relatively unaffected. (Subsequent androgen replacement injections produced a differential facilitation of posterior hypothalamic self-stimulation rate.) Conversely, food deprivation increases lateral hypothalamic but not posterior hypothalamic self-stimulation rate. This experiment thus shows that at one site stimulation produces sex-related reward, and at another site hunger-related (i.e. food) reward. Olds (1958, 1961) has also suggested that androgens and hunger have differential effects on self-stimulation at different sites.

There is now direct evidence that brain-stimulation reward can produce an effect equivalent to a specific natural reward (experiments of E. T. Rolls, M. J. Burton, and S. G. Shaw—see Rolls, 1974). Recordings from single hypothalamic cells were made in the squirrel monkey during self-stimulation. It was found that some single units fired when stimulation was applied to reward sites in, for example, the orbitofrontal cortex, the nucleus accumbens, and the lateral hypothalamus (see, for example, Fig. 20). The latency to firing was typically between 1 and 30 ms. The function of these units in natural behaviour was then analysed. Because damage to this region of the brain produces aphagia and adipsia (lack of eating and drinking—see, for example, Grossman, 1973, ch. 9), the firing of the units during eating and drinking was measured. Some of the neurones altered their firing while food was in the mouth, and others while water was in the mouth. (Some other units fired as soon as the hungry monkey saw food but not when he saw an inedible or aversive object.) An example of one of the units is shown in Fig. 12. The unit fired while water was in the mouth of the hungry, thirsty monkey, but not while

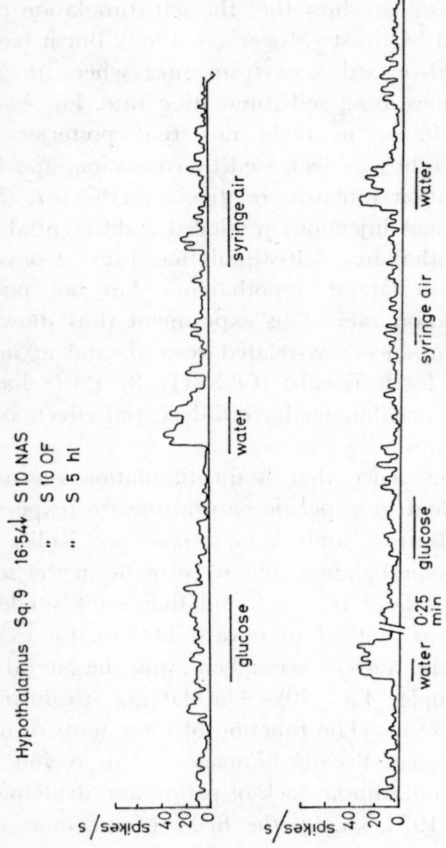

FIG. 12. Unit in the hypothalamus of the squirrel monkey trans-synaptically (S) activated during self-stimulation of the nucleus accumbens (NAS), orbitofrontal cortex (OF), and lateral hypothalamus (hl) with the latencies (in ms) shown. The unit fired more when the hungry and thirsty monkey drank water, but little change was seen when 5% glucose (or isotonic saline) was drunk, or when air was blown on the mouth from a syringe (this latter is aversive). The unit fired most rapidly while the water was in the mouth. Time scale: the stimuli were presented for approximately 20 s.

isotonic glucose or isotonic saline was in the mouth, and not when an empty syringe (syringe, air) was held to the lips. Thus the unit appeared to tell the hungry, thirsty monkey that a specific natural reward—water—was available. In that brain-stimulation reward of at least some sites also activated this type of unit, it can be said that brain-stimulation reward mimics the effect of a specific natural reward. A sufficient condition for electrical stimulation of the brain to provide reward may be that it mimics the effect of a specific natural reward.

The conclusion which follows from the observations that self-stimulation can be modulated by deprivation, and that brain-stimulation reward can mimic the effect of a natural reward, is that intracranial self-stimulation may occur because it mimics the effects of natural rewards.

2.1.3. *Further evidence on hypothalamic reward cells*

Several other observations suggest that the hypothalamic cells described above are involved in natural reward. Firstly, lesions of the lateral hypothalamus lead to a lack of eating and drinking. This is clearly consistent with the view that some hypothalamic cells maintain and guide eating and drinking, i.e. produce reward.

Secondly, food and water are provided mainly by sensory input, and (see above) some hypothalamic cells respond to this sensory input. That oropharyngeal sensory input, in particular taste and smell, rather than gastric factors (e.g. stomach distension) or post-absorptional factors, mediate reward is shown by the observations that animals with oesophageal fistulae eat and drink at least as much as intact animals—food and water intake is maintained (see Grossman, 1973). (In oesophageal animals food is tasted, smelled, and swallowed normally, but passes out of the animal through a tube attached to the oesophagus.) Further evidence that oropharyngeal factors normally guide and maintain intake is that with intragastric self-administration in the rat the normal preference for low concentrations and aversion for high concentrations of sweet substances are not found (there is almost no guiding of intake), relatively large volumes (e.g. 0.5 ml for intragastric administration as compared with 0.01 ml when tasted normally) of food must be given for each response, and the rats transfer only with difficulty to intragastric self-administration (there is only poor maintenance of intake) (Epstein, 1967). Similarly, rats require relatively large volumes (e.g. 0.5 ml) of water per bar-press if intake is to be maintained with intravenous self-administration and learn the task only with difficulty (Rowland, 1973). A further point is that animals (and man) will work to obtain saccharin solution, which has a sweet taste but is metabolically inert. Although sweet substances placed on the tongue may lead to a rise of blood glucose (Nicolaidis, 1969) and this could play a part in reward, the taste is still involved even in this scheme. These points provide clear evidence that oropharyngeal sensory input normally provides reward, i.e. guides and maintains intake. The advantage of this is that the animal rapidly gains information about possible nutrients (within 1 s) and can try to obtain more or less of the substance instead of having to delay until post-ingestional effects become apparent minutes later. (Learning without immediate reinforcement is generally poor.) The finding that lateral hypothalamic lesions also lead to a neglect of sensory input so that lesioned rats do not orient to food or objects presented visually or to touch and even smell (Marshall *et al.*, 1971) is also consistent with the finding that hypothalamic cells respond to sensory input and that this sensory input is involved in reward. We have observed that the effects of visual stimuli on these hypothalamic cells can be modified by experience, so that a white syringe containing glucose may elicit firing while a black syringe containing water may not. The rewarding properties of visual stimuli appear to be represented at the hypothalamic level.

Thirdly, the response of some of the activated hypothalamic cells to natural reward, e.g. water, became less after the animal had drunk water and was less thirsty. Hypothalamic cells whose response to the smell of food was modulated by hunger have been described by Le Magnen and Vincent (1973). The cells responded to food but not to non-food odours, and only responded if the monkey was hungry. This type of hypothalamic cell then has the appropriate properties for signalling food reward—the monkey responds to the smell of food as a reward only if he is hungry. As this type of cell is activated by brain-stimulation reward of at least some sites (see above) and the brain-stimulation reward can also be modulated by drives such as hunger and thirst, it seems even more likely that activation of this type of cell can mediate brain-stimulation reward. It is also, of course, consistent that hypothalamic cells which are thought to mediate reward are activated by brain-stimulation reward.

The factors which modulate reward can be specified in some detail in the case of food reward. These are the factors which control food intake and signal hunger. A volume signal, mediated perhaps by stomach distension, is of partial importance in the control of meal size. In the dog, some eating will still occur if a stomach preload of 175% of the expected food is given (Janowitz and Hollander, 1953), and subsequent intake is not affected unless more than 20% of a normal meal is given as a gastric pre-load (Janowitz and Grossman, 1949). In man, volume signals appear to be relatively more important in switching off food intake in that after a rapid oral pre-load of half the normal intake, the volume of food subsequently eaten makes up the normal intake rather precisely (Spiegel, 1973; see Fig. 13). The regulation was by volume and was not caloric in that the caloric density of the pre-load did not affect intake (Fig. 13). Volume is further seen to be of importance in that in similar experiments in man after intragastric pre-loads precise regulation is also found (Jordan, 1969; Walike *et al.*, 1969). In these experiments, the subjects gave ratings of their degree of hunger and fullness. Hunger ratings were depressed, and fullness ratings increased, by the pre-loads. Thus a volume factor depresses subsequent intake and reduces feelings of hunger. A similar volume factor appears to decrease the reward value of lateral hypothalamic stimulation in the rat, as described above.

In the rat, altered glucose levels (or utilization) affect both the reward value of lateral hypothalamic stimulation and feeding (see above). Similarly, the subjective rating of a sweet taste or of a smell of food is moved away from +2 (very pleasant) towards −2 (very unpleasant) by loads of 100 g of glucose in man (Cabanac, 1971; Cabanac *et al.*, 1971; see Fig. 14). The precise nature of the stimulus is not quite clear, as stomach loads of saccharin can produce a similar effect (Wooley *et al.*, 1972). The release of glucose during eating occurs remarkably soon after the onset of a meal—within 2–5 min in the rat (Steffens, 1970; see Fig. 15). It is of interest that the glucose must be released from stores in the liver in that radioactively labelled

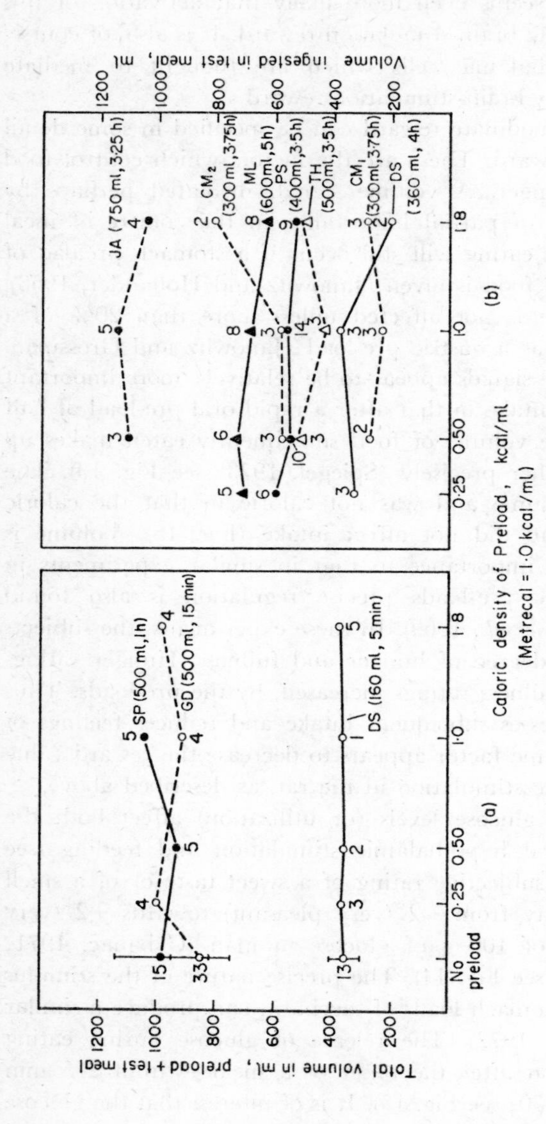

FIG. 13. Control of food intake by a volume factor in man. The mean volume of liquid diet ingested (test meal plus pre-load, with the pre-load volume half that taken usually) as a function of the caloric density of pre-loads given less than 1 h previously (a) and 3–5 h previously (b) Subjects compensated fairly accurately for the volume but not for the caloric content of the pre-load. (Further details are given by Spiegel, 1973.) (From T. A. Spiegel, Caloric regulation of food intake in man, *J. Comp. Physiol. Psychol.* **84**, 24–37, 1973. Copyright by the American Psychological Association, and reproduced by permission.)

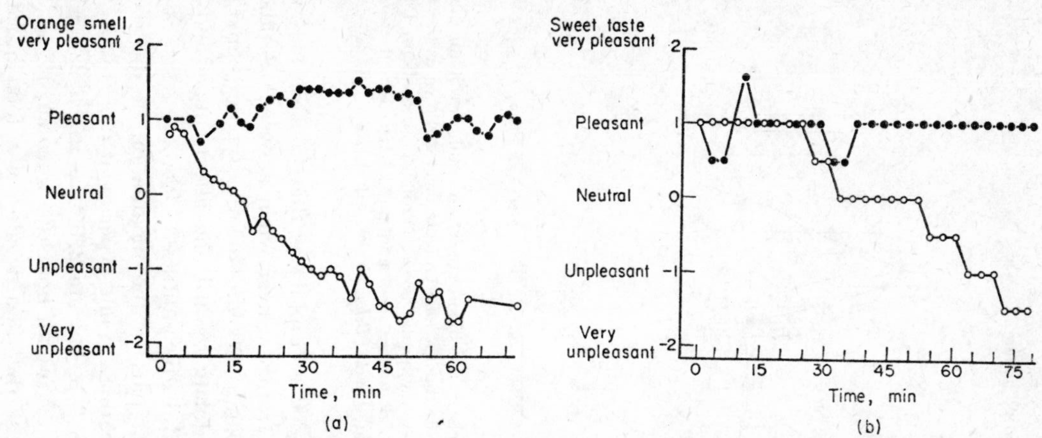

FIG. 14. Modulation of the pleasantness of taste and smell in man by a hunger signal. (a) Affective responses of the one fasting subject when given an olfactory stimulus related to feeding. The same stimulus was repeated throughout the experiment. *Open circles:* the subject ingested 100 g glucose in water at the time shown by the arrow. *Closed circles:* control, no glucose administered. (b) Affective responses given by one fasting subject to a sweet taste. The same stimulus was repeated throughout the experiment. *Open circles:* the subject swallowed the test samples (10 g of sucrose in 50 ml). *Closed circles:* the subject expectorated the samples after tasting. (From M. Cabanac, Physiological role of pleasure, *Science*, **173,** 1103–7, 1971. Copyright by the American Association for the Advancement of Science, and reproduced by permission.)

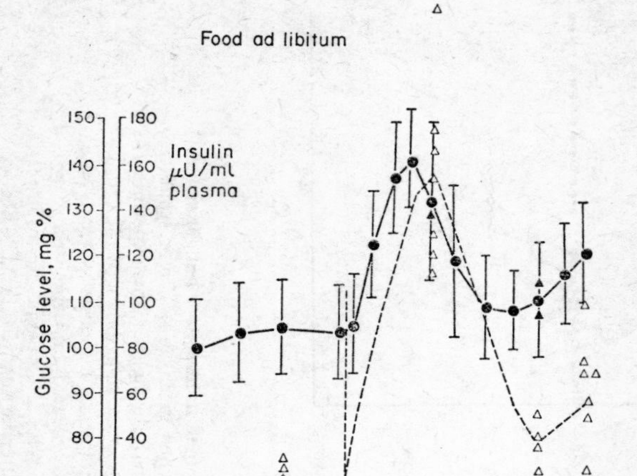

Fig. 15. Glucose and insulin levels in the blood rise within minutes of the start of a meal in the rat. The 8 rats had food *ad libitum.* Glucose level is shown as the mean ±SD. Individual points are shown for insulin levels. (From A. B. Steffens, Plasma insulin content in relation to blood glucose level and meal pattern in the normal and hypothalamic hyperphagic rat, *Physiol. Behav.* **5,** 147–51, 1970.)

food does not reach the blood as rapidly as this. Glucose may even be released into the blood from the liver when a sweet substance is placed on the tongue (Nicolaidis, 1969). Thus this type of signal may be sufficiently rapid to modulate reward and affect intake and subjective ratings of hunger and the pleasantness of food.

Taste factors also play a part in the modulation of the reward value of food, and thus of eating. In one experiment rats ate a certain amount of standard diet in 2 h, but ate 270% as much if the food were given a different taste every half hour (Le Magnen, 1971). Thus prolonged taste may decrease the reward value of a food in satiety, or a new taste may increase the reward value of food. This is seen also in the salted-nut phenomenon (Hebb, 1949) in which if one is given a nut, more are desired. Similarly, the rate of feeding increases in the first few minutes of a meal in rats (Le Magnen, 1971). These are examples of incentive motivation.

The experiments described above show that the reward value of food, and hence the amount of food eaten and the subjective rating of the pleasantness of food, are modulated by factors such as stomach distension, carbohydrate metabolism, and body weight. These factors signal the level of hunger, and it appears that the hunger modulates reward. (In psychological terms we may say that drive modulates reward.) One point which follows from the experiments on self-stimulation is that this modulation occurs in, or near, the lateral hypothalamus, as this is a site at which brain-stimulation reward is modulated by hunger. The principle of operation of these processes is illustrated in Fig. 16A. Sensory input from, for example, taste or smell reaches the hypothalamus. The effects of this sensory input on reward neurones are gated by hunger, so that the sensory input influences reward neurones only if the animal is hungry. The cell bodies of the hunger neurones need not be present in the lateral hypothalamus, e.g. destruction of the lateral preoptic area (or the lateral hypothalamus) abolishes eating induced by a decrease of glucose utilization (Miselis and Epstein, 1971). Self-stimulation could occur if the neurones which carry the sensory input corresponding to food are excited by electrical stimulation. Modulation of self-stimulation by hunger in other brain areas could be due to an independent mechanism, or could be because areas related to this lateral hypothalamic system are activated. The evidence described above suggests that there are similar mechanisms in or related to the hypothalamus which regulate thirst and influence sexual behaviour. It is not clear whether the other more general reward mechanisms include a modulatory system of this nature (see below).

The operation of this type of mechanism for the control of food intake in which the sensory stimuli associated with food intake produce reward in hypothalamic neurones only if the animal is hungry is illustrated in some experiments on the monkey (E. T. Rolls and M. J. Burton, Fig. 16A). In recordings from single hypothalamic neurones it was found that the taste of food (e.g. a glucose solution) or the sight of food (e.g. the sight of a peanut, or of a syringe from

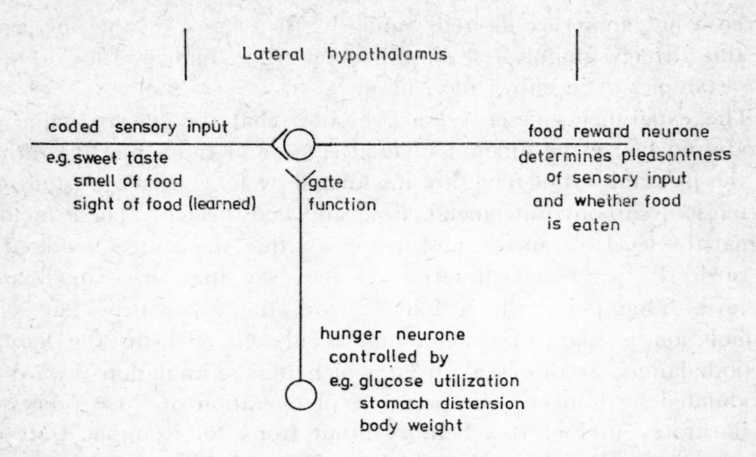

FIG. 16A. Control system for feeding consistent with experiments on brain-stimulation reward (see text). The effects of coded sensory input on lateral hypothalamic reward neurones are gated by neurones which signal hunger. The input is coded beyond the receptor level so that, for example, the smell of food but not of non-food objects is signalled, or, for example, the learned sight of a food but not of a non-food object is signalled. The gating (which could be pre- or post-synaptic) by "hunger" neurones is performed on the basis of factors such as glucose utilization, stomach distension, and body weight. Activity in the food-reward neurones determines the pleasantness of the sensory input and whether feeding will continue. (The initiation of feeding depends on activity in the "hunger" neurones.) Self-stimulation of the lateral hypothalamus could occur because of excitation of either the coded sensory input neurones, in which case modulation by hunger would occur, or because of excitation of the food-reward neurones.

which the animal was fed glucose) only affected the firing rate of the neurones if the monkey was hungry. Hunger was manipulated by feeding the animal glucose. In the case of the visual hypothalamic neurones learning occurred, so that the neurones only fired to a visual stimulus if it was associated with food intake. It was even found that if the animal was no longer fed from, for example, a black syringe, then the neurones no longer fired to the black syringe. The visual hypothalamic neurones may thus be important in enabling a hungry animal to recognize food visually, and may help the hungry animal to trace food by following visual stimuli previously associated with food intake. In the same experiments it was found that these sensory hypothalamic neurones were activated with short latencies (e.g. 5–30

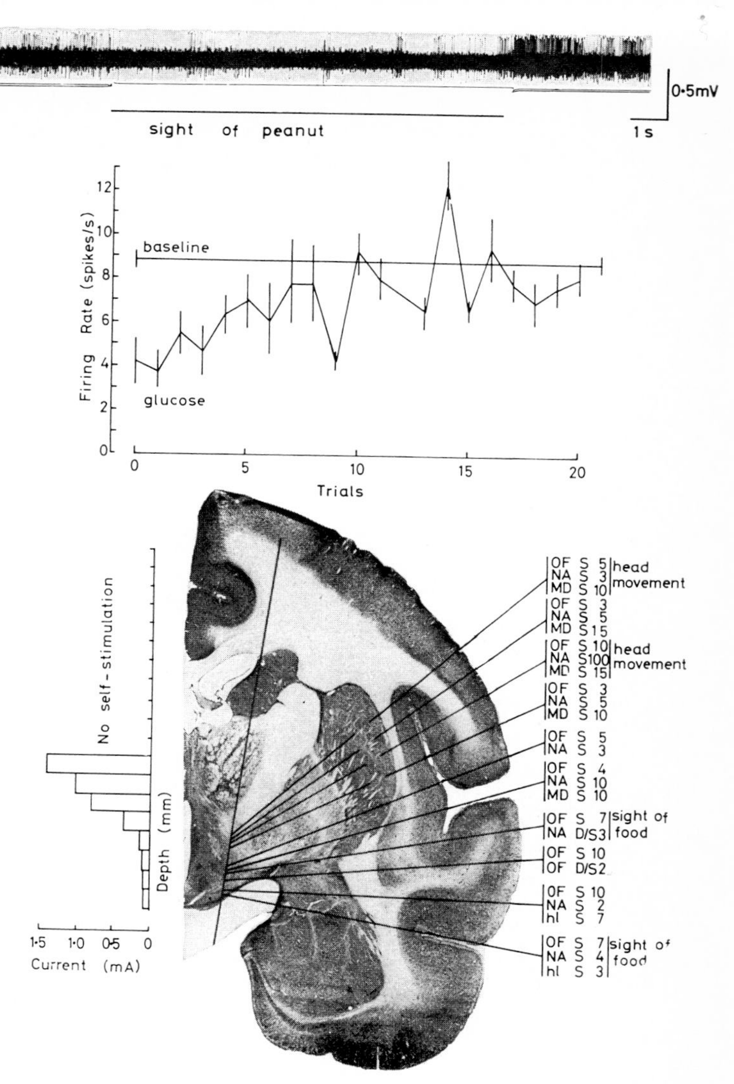

Fig. 16b. Hypothalamic units affected by both brain-stimulation reward and natural rewards. *Top.* This lateral hypothalamic unit, which was activated in self-stimulation (for example, see Fig. 19), almost stopped firing when the monkey saw a peanut. *Middle.* A similar lateral hypothalamic unit which showed a reduction of its firing rate from the spontaneous baseline rate of 9 spikes/s to a rate of 4 spikes/s when a monkey was shown a black syringe which he had learnt contained glucose. The monkey was fed some of the glucose on every trial, and as he became less hungry (trials 0–15) the neurone responded less to the sight of food. *Bottom.* Units activated by both brain-stimulation reward and the sight of food were found at the lower end of this track, in the hypothalamus. Units higher up the track in the globus pallidus were activated by brain-stimulation reward and also by head movements. Self-stimulation through the recording microelectrode occurred in the hypothalamic region, where cells fired to the sight of food, and the self-stimulation here was more intense if the animal was hungry. (Experiments of E. T. Rolls and M. J. Burton.)

ms) by brain-stimulation reward of a number of sites. It was further found that self-stimulation would occur if the stimulation was applied through the recording microelectrode as long as the microelectrode was in the region of these hypothalamic cells. We also showed that the self-stimulation with the microelectrode in the region of the lateral hypothalamus was best when the animal was hungry, and needed a higher current when the animal was satiated. Examples of this type of unit are shown in Fig. 16B. Taken together with the evidence that lesions in this region disrupt food intake, the experiments provide strong support for the hypothesis that food intake occurs when hypothalamic reward neurones are gated by hunger to respond to sensory inputs such as the sight, smell, and taste of food.

2.1.4. *Other types of reward*

At some sites, brain-stimulation reward does not appear to be modulated by a drive such as hunger or thirst. For example, Hoebel (1967, 1968) demonstrated that feeding decreased lateral hypothalamic but had rather little effect on septal self-stimulation rate (see Fig. 10). It was also shown that feeding made rats prefer the septal stimulation. If brain-stimulation reward at a particular site cannot be shown to be modulated by any factor, then a possibility is that stimulation at that site produces a general reward equivalent to general pleasure. Clearly it is difficult to analyse this further in animals. The question can be answered by asking a human subject what he feels during the stimulation.

2.2. EVIDENCE FROM STUDIES ON MAN

Human patients are able to report the effects of electrical stimulation of the brain which has been performed during the investigation or treatment of tumours, epilepsy, schizophrenia, or Parkinson's disease. Emotional feelings such as anger, joy, pleasure, and sexual excitement are almost never evoked by electrical stimulation of the (easily exposed) cerebral cortex (Penfield and Jasper, 1954). Sensory and motor changes were produced by the stimulation. This is consistent with the finding that self-stimulation of the cerebral cortex is not typically found in animals (see Chapter 3). However, electrical stimulation of some regions within the human brain can induce pleasurable manifestations, demonstrated by the spontaneous verbal reports of patients, their facial expression and general behaviour, and their desire to repeat the experience (see Delgado, 1969; Sem-Jacobsen, 1968). In an early study in twenty-three schizophrenic patients electrical stimulation of the septal region produced alertness sometimes accompanied by an increase in verbal output, euphoria, or pleasure (Heath, 1954). Self-stimulation in man has been formally tested by giving patients a button to press to deliver short trains of electrical stimulation to their brains. In two patients self-stimulation of, for example, the amygdala, caudate nucleus, and mesencephalic tegmentum occurred at rates of 250–400 presses/h for 0.5 s trains (Heath, 1963; Bishop *et al.*, 1963). The patients reported that the stimulation at these sites felt "good" and "pleasurable". Self-stimulation of the septal region (defined in these studies as the region of the nucleus accumbens septi and diagonal band of Broca) also occurred, and was accompanied by sexual thoughts. One patient felt as if he were building up to a sexual orgasm. In two patients in whom EEG recordings were made (Heath, 1972), spike and slow waves were prominent in this septal region during orgasm. The injection of acetylcholine (400 μg) into the septal region was also capable of producing reward related to sexual behaviour. Sem-Jacobsen (1968) has also found intracranial self-stimulation in man. Delgado (Higgins *et al.*, 1956; Delgado, 1960; Delgado and Hamlin, 1960; see Delgado, 1969) has found in three patients with psychomotor epilepsy that stimulation in the temporal lobe was capable of producing relaxation and pleasure accompanied by different effects such as tingling in one side of the body or sexual thoughts. Sem-Jacobsen (1968) has described the effects of stimulation at 2659 sites in eighty-two patients with schizophrenia or Parkinson's disease. Responses were obtained from 1594 electrodes. Changes in mood were elicited from 643 electrodes in sixty-six patients. Most of these electrodes were in the anterior part of the brain. At 360 points the patients became relaxed, at ease, had a feeling of well-being, and/or were a little sleepy (classed as Positive I). At thirty-one sites the patients, in addition, showed enjoyment, frequently smiled, and might want more stimulation (Positive II). At eight sites (in seven patients) the patients laughed out loud, enjoyed themselves, positively liked the stimulation, and wanted more (Positive III). In addition, pleasant smells were evoked at nine sites and unpleasant at six, mostly with electrodes in the frontal lobes. Pleasant tastes were evoked at three sites, and unpleasant at one, with electrodes near the tip of the temporal pole. Sexual responses were obtained from two electrodes. During investigations of two patients with temporal lobe epilepsy it was found that electrical stimulation of the amygdala could elicit emotional responses (see Fig. 17) (Ervin *et al.*, 1969; Mark and Ervin, 1970; Mark *et al.*, 1972). In these patients it was also observed that the violent behaviour of these patients occurred shortly after large EEG waves had started in the temporal lobe, and that the patients were more manageable after damage to this area (for discussion see Valenstein, 1974). A feature of the emotional feelings described in man is that they may outlast the stimulation for minutes or hours (see, for example, Ervin *et al.*, 1969).

These observations on man can help us to understand brain-stimulation reward. They show that electrical stimulation of the brain can be pleasant. This conclusion is consistent with the conclusion from studies on animals that animals will work to obtain electrical stimulation of the brain because they like it. In man it appears that rewarding stimulation at many sites produces a general type of pleasure, although at some sites specific reward is produced. For example, it appears that septal stimulation can produce sex-related reward. In animals, food, water, and sex-related reward are commonly described. This may be because hypothalamic sites have been studied more in animals, and/or because a greater part of the human brain is concerned with other types of reward.

Although intracranial self-stimulation in man is usually performed because the patients like the stimulation, this is sometimes not the case. For example, one patient was almost able to recall a memory during stimulation of the centromedian thalamus but could not quite grasp the memory. He self-stimulated frequently in an endeavour to bring the elusive memory into clear focus (Heath, 1963). Sem-Jacobsen (1968, p. 54) has also indicated that some patients may self-stimulate because of a desire to co-operate rather than because of reward. Also, facial expression cannot be taken as a perfect guide to

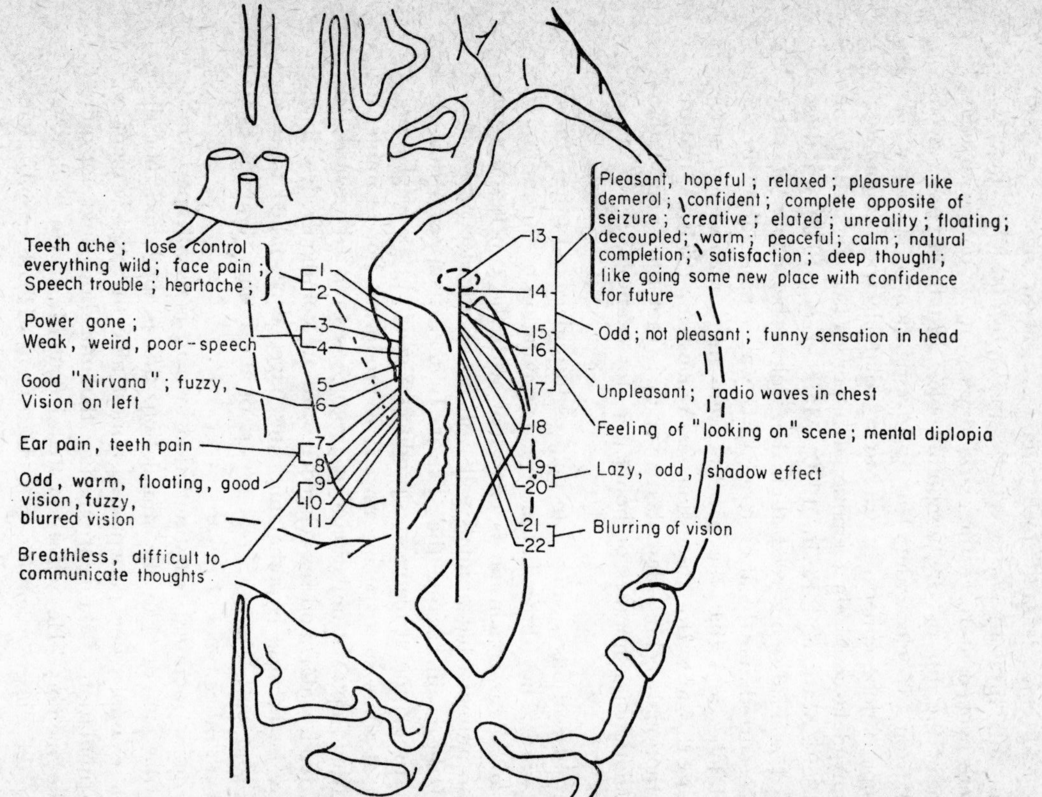

FIG. 17. Summary of subjective states evoked by electrical stimulation in the human temporal lobe on a diagrammatic reconstruction of the stimulating points. Most of the sensations occurred during the electrical stimulation, but the pleasure and euphoric effect could outlast the stimulation for minutes or hours. (From J. R. Stevens, V. H. Mark, F. Ervin, P. Pacheco, and K. Suematsu. Deep temporal stimulation in man, *Archs. Neurol.* **21**, 157–69, 1969. Copyright the American Medical Association, and reproduced by permission.)

pleasure, as in one of his patients stimulation elicited tickling and laughing with no enjoyment at all. Fortunately, humans can describe the effects of the stimulation. It appears that intracranial self-stimulation in man does usually occur because the stimulation is pleasant.

2.3. EATING, DRINKING, AND SEXUAL BEHAVIOUR PRODUCED BY STIMULATION OF SOME REWARD SITES

If continuous electrical stimulation is applied to some hypothalamic sites in rats, eating, drinking, or sexual behaviour may occur. A rat typically starts to eat or drink within a few seconds of the onset of stimulation, and continues to eat or drink as long as the stimulation is on. This type of behaviour is called "stimulus-bound". Margules and Olds (1962) noted that self-stimulation would occur at most of the lateral hypothalamic sites at which stimulus-bound eating was elicited. What then is the relation between the brain-stimulation reward and the stimulus-bound behaviour?

The motivation produced in stimulus-bound behaviour is sometimes non-specific in that stimulus-bound feeding may change to stimulus-bound drinking if the food is removed and replaced with water (Valenstein *et al.*, 1970). However, there is evidence that specific drives can be elicited, and that the non-specificity often observed may be due to the simultaneous stimulation of several of these different specific systems. Sexual behaviour such as copulation and ejaculation in the male rat can be elicited by stimulation of the posterior hypothalamus (Herberg, 1963; Caggiula and Hoebel, 1966; Caggiula, 1967, 1970) and lateral preoptic area (Madlafousek *et al.*, 1970) and may produce no facilitation of feeding and drinking (Madlafousek *et al.*, 1970). Eating or drinking may be elicited by stimulation of areas in or near the lateral hypothalamus. Olds *et al.* (1971) claim that stimulation in an anterior dorsolateral hypothalamic area encroaching on the zona incerta elicits drinking, that stimulation in a middle dorsolateral hypothalamic area also encroaching on the zona incerta elicits eating, and that self-stimulation alone without eating and drinking is produced at electrodes in the hypothalamus near the medial forebrain bundle. In their attempts to localize specific effects in the hypothalamus, Olds *et al.* (1971) used nichrome wire with a small (62 μm) diameter. Huang and Mogenson (1972) also found specificity in that stimulation sites in the zona incerta in the region of pathways to the thalamus and subcommissural organ elicited drinking, and stimulation of lateral hypothalamic sites from which fibre degeneration spread to the medial forebrain bundle elicited feeding. In both studies both eating and drinking were elicited from sites dorsolateral to the fornix. Further evidence for specificity is provided by Caggiula (1967), who found in the same rat that lateral hypothalamic stimulation elicited feeding and posterior hypothalamic stimulation elicited copulatory behaviour. Thus, feeding, drinking, and sexual behaviour can apparently be elicited specifically, as can water reward and food reward (Gallistel and Beagley, 1971; see above).

Evidence that stimulus-bound behaviour and self-stimulation are closely related at some sites comes from the finding that only at stimulus-bound feeding sites did food deprivation increase self-stimulation rate (Goldstein *et al.*, 1970). Thus food-reward neurones may be present at sites at which stimulation elicits feeding. Similarly, castration in male rats decreased self-stimulation at a posterior hypothalamic site from which copulation was induced (Caggiula, 1967, 1970). The castration did not have a similar effect on lateral hypothalamic self-stimulation rate. Thus sex-reward neurones may be present at sites at which stimulation elicits sexual behaviour, but not at other sites. This is an interesting relation which suggests that areas at which stimulation elicits a motivational behaviour contain reward neurones for that type of behaviour. At present it is not clear whether this is a necessary relation. It may not be if the observation of Olds *et al.* (1971) that eating or drinking can be elicited at non-self-stimulation sites is correct.

One explanation of how stimulus-bound motivational behaviour is produced is that the drive neurones which signal hunger, thirst, etc., and modulate reward neurones (see Fig. 16) are activated by the stimulation. On this view, stimulus-bound motivational behaviour is like normal hunger or thirst, or sometimes like both combined depending on the site of the stimulation electrode. Another tentative explanation is that the neurones which mediate reward are activated, and that this produces incentive motivation which together with other effects produced by the stimulation (e.g. arousal, see section 3.7.4) elicits eating or drinking. Incentive motivation is the increased intensity of goal-directed behaviour which follows the presentation of a reward (see p. 26). For example, the presentation of a salted nut increases one's desire for salted nuts (Hebb, 1949). It is seen, for example, as increased bar-pressing for a food reward produced in rats by the intraoral injection of a small quantity of chocolate milk just before the bar-pressing (Panksepp and Trowill, 1967b).

One observation which has caused confusion is that when the stimulating current is set just below the threshold for self-stimulation, rats will self-stimulate if water (Mendelson, 1970) or food (Coons and Cruce, 1968) is also delivered for every response. This may well only indicate summation of natural reward with brain-stimulation reward.

2.4. SUMMARY

Brain-stimulation reward at some sites mimics a natural reward such as food in that, for example, animals made hungry self-stimulate faster for lateral hypothalamic stimulation. The reward can be like a specific natural reward in that at some sites it can mimic food and at other sites water. Recordings from single cells in the hypothalamus show that some cells excited by naturally rewarding stimuli such as the taste of glucose or water are also excited by brain-stimulation reward at some sites. It is proposed that a sufficient explanation for brain-stimulation reward is that it excites pathways concerned with natural rewards.

The hypothalamic cells which respond to natural rewarding stimuli only respond fully if the animal is hungry, thirsty, etc. Thus hunger (signalled by internal variables such as the level of glucose in the blood and body weight) modulates the effects of sensory stimuli on hypothalamic cells: drive modulates reward. It is probably by this mechanism that brain-stimulation reward at some sites can be modulated by hunger, etc., and that the pleasantness of sensory stimuli to humans is modulated by hunger and other drives. These experiments with brain-stimulation reward emphasize the point that motivational behaviour (such as eating and drinking) is maintained (rewarded, reinforced) by sensory stimuli (such as the smell or taste of food) which have a reward effect on hypothalamic cells gated by need (or drive, e.g. hunger and thirst).

At other sites brain-stimulation reward occurs because of the excitation of pathways concerned with more general emotional states: human subjects report that pleasure or happiness is produced by stimulation at some sites.

CHAPTER 3

THE NEURAL BASIS OF BRAIN-STIMULATION REWARD

A STUDY of the neural basis of intracranial self-stimulation may clarify the nature of brain-stimulation reward. It may be further useful in several ways. Firstly, it may indicate brain regions concerned with natural reward. For example, as described above, the control of eating probably occurs by adjustment of the reward value of food, and this reward control system can be tapped and analysed with electrical stimulation. Similarly, reward and punishment are closely related to emotional behaviour, and the neural basis of emotional behaviour can be analysed with brain-stimulation reward. Secondly, it may clarify the basis of abnormal human emotional behaviour some types of which Stein (1971) has suggested are due to malfunction of the reward system. Thirdly, it may clarify the relief from intractable pain produced by stimulation of some rhinencephalic reward sites in man (Brady, 1961; Heath, 1964). Fourthly, the hypothalamus and limbic structures contain potent reward sites, and analysis of the neural basis of reward may clarify the function of these regions of the brain. Fifthly, reward is often an important factor in learning, and analysis of brain-stimulation reward may clarify the neural basis of learning.

3.1. ANALYSIS OF THE NEURAL BASIS OF BRAIN-STIMULATION REWARD

The neural basis of brain-stimulation reward can be analysed by determining which parts of the brain support brain-stimulation reward, which parts if lesioned or damaged affect brain-stimulation reward, the course of fibres from self-stimulation with the use of fibre degeneration methods, and which neurones are active during brain-stimulation reward. Pharmacological methods can also be used to analyse the neurotransmission involved in brain-stimulation reward.

Much of the evidence on the neural basis of brain-stimulation reward comes from studies in animals. Comparable data are not in general, of course, available for man, and even sites where stimulation elicits pleasure in man cannot be described in detail because the brains have not come to post mortem (e.g. Sem-Jacobsen, 1968). Nevertheless, some generalizations about the neural basis of reward in man can be made from studies on man in which the stereotaxic coordinates of the stimulation site are known. Stimulation of most areas of the neocortex in man does not elicit pleasure, but stimulation in the depths of the brain in parts of the hypothalamus and limbic structures such as the amygdala can elicit reward (Sem-Jacobsen, 1968; Delgado, 1969; see above). A similar situation occurs in animals (see below). Further, in man the effects of electrical and chemical stimulation of the septal region, and EEG recording, suggest that this area may be involved in sex-related reward. In animals, this same general region of the diagonal band of Broca and the nucleus accumbens septi, which is just rostral to the anterior preoptic area, is involved in sexual behaviour (Madlafousek *et al.*, 1970; see above; and also Grossman, 1973).

Diagrammatic representations of some limbic–hypothalamic pathways are shown in Figs. 18A, 18B. Descriptions of hypothalamic connections can be found in Zeman and Innes (1963), Raisman (1966), and Brodal (1969). References to particular papers are given below where appropriate.

3.2. BRAIN-STIMULATION REWARD SITES

Work before 1965 has been summarized by Olds and Olds (1965). One group of sites extends along the medial forebrain bundle (MFB) from the region of the diagonal band of Broca to the ventral tegmental area of Tsai, which is just lateral to the interpeduncular nucleus in the midbrain. Olds *et al.* (1971) have reported that for the hypothalamus self-stimulation can be obtained near the MFB, while

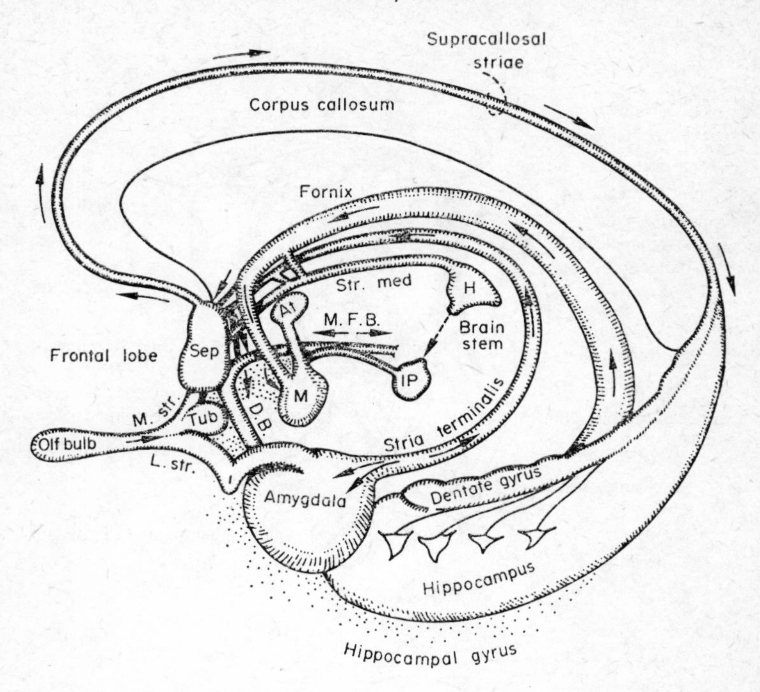

Fig. 18a. A representation of some limbic connections. L.str., lateral olfactory tract; M.str., medial olfactory tract; Tub, olfactory tubercle; Sep, septal area; D.B., diagonal band of Broca; M.F.B., medial forebrain bundle; IP, interpeduncular nucleus (in ventral tegmental area); H, habenular nucleus; Str.med., stria medullaris; M, mamillary body; At, anterior nucleus of the thalamus. (After P. D. MacLean.)

feeding or drinking is produced by stimulation more dorsolaterally. A second group of sites is in the rhinencephalon. In the rat, rhinencephalic self-stimulation has been reported for parts of the septal nuclei (Olds *et al.*, 1960); nucleus accumbens septi (Routtenberg and Huang, 1968); amygdala (Wurtz and Olds, 1963; Valenstein and Valenstein, 1964; Hodos, 1965); hippocampus (Ursin *et al.*, 1966; Milgram, 1969); entorhinal, retrosplenial, and cingulate juxtallocortex (Stein and Ray, 1959; Brady and Conrad, 1960); and the caudate nucleus (Olds, 1960).

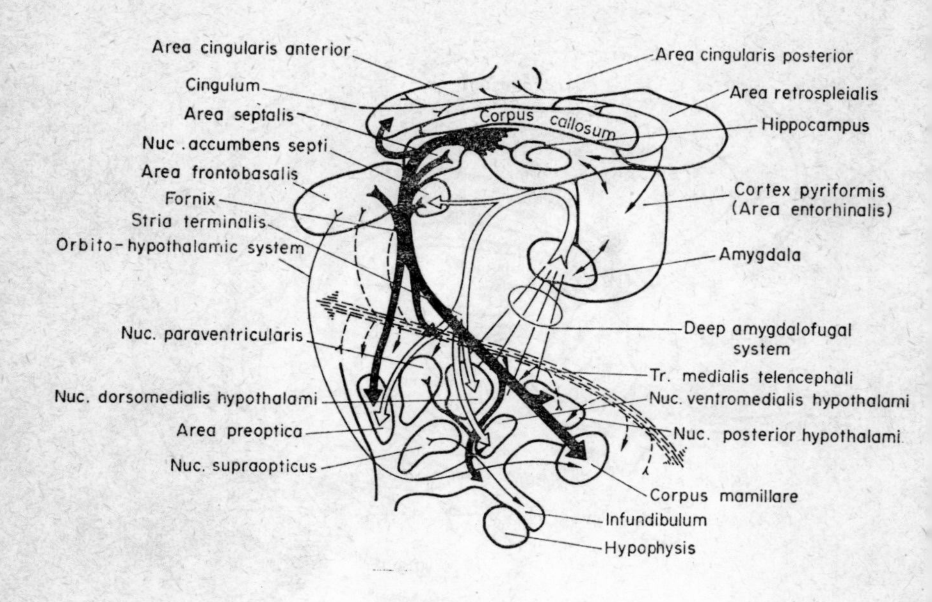

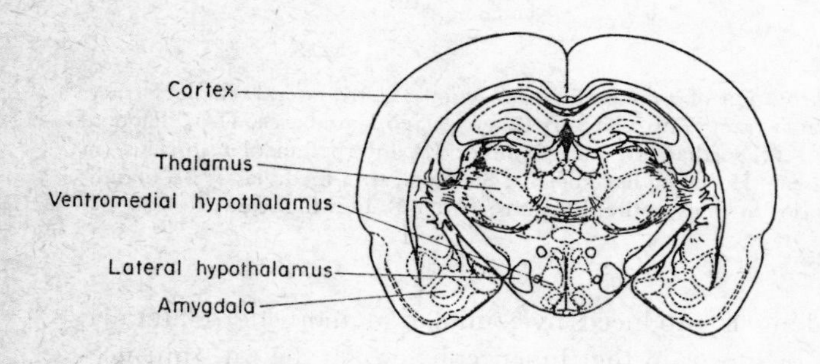

FIG. 18B. *Upper:* Some pathways related to the hypothalamus. The area frontobasalis corresponds to the orbitofrontal cortex. The tractus medialis telencephali is the medial forebrain bundle. The deep amygdalofugal system corresponds to the ventral amygdalofugal pathway, the ventral pathway, and the direct amygdalofugal pathway, and is bidirectional. (From W. Zeman and J. R. M. Innes, *Craigie's Neuroanatomy of the Rat*, Academic Press, New York, 1963.) *Lower.* Coronal section of the rat brain at the hypothalamic level.

Major differences between MFB and rhinencephalic self-stimulation have been summarized by Olds and Olds (1965). MFB self-stimulation occurs at much higher rates than rhinencephalic self-stimulation—under similar conditions rates may be 10,000 vs. 500 responses/h. This does not necessarily mean that MFB stimulation is more rewarding than rhinencephalic stimulation, because response rate is not necessarily a good indication of reward measured by preference. For example, Hodos and Valenstein (1962) found that although rats preferred medium intensity septal stimulation to low intensity hypothalamic stimulation, the rate of bar-pressing was lower for the septal stimulation. One factor which contributes to the high MFB self-stimulation rates is arousal which the stimulation produces in addition to its rewarding effect (see section 3.7.4). Rhinencephalic self-stimulation may show "satiation" and stop for the day after several thousand bar-presses, while MFB animals continue until exhausted (although MFB animals requiring priming may show some decline in rate; Kent and Grossman, 1969). MFB reward is accompanied by hyperactivity, whereas at least during intracranial stimulation, rhinencephalic rewarding stimulation is accompanied by hypoactivity. Pain or anxiety relief in man is obtained from stimulation of some rhinencephalic reward sites, and no relief is obtained from some MFB sites (Brady, 1961, but see section 5.5). This situation may also occur in the rat, in which rewarding rhinencephalic (septal) stimulation suppresses the aversive effects of tegmental stimulation (Routtenberg and Olds, 1963), but rewarding MFB stimulation augments the behavioural response to the aversive stimulation (Olds and Olds, 1962; Stein, 1965). It must be noted though that hypothalamic stimulation can attenuate the aversive properties of peripheral shock (e.g. Cox and Valenstein, 1965). Further, Gardner and Malmo (1969) found that although rewarding septal stimulation reduced escape from midbrain stimulation, the same septal stimulation facilitated escape from foot shock. Finally, the eating, drinking, or sexual activity which may be produced by stimulation of MFB reward sites is stimulus-bound (for review see Valenstein *et al.*, 1969), while feeding produced by stimulation of some hippocampal reward sites occurs for 4–30 s after a 1 s period of stimulation, and may be regarded as a rebound phenomenon (e.g. Milgram, 1969).

Self-stimulation may also occur outside these areas. Cooper and Taylor (1967) found self-stimulation in the thalamic reticular system, and also in the central grey of the midbrain, where it developed over several days. Routtenberg and Malsbury (1969) found self-stimulation of certain brain-stem sites, in particular in or near the substantia nigra, brachium conjunctivum and rubrospinal tract, in addition to the ventral tegmental area of Tsai, but not of the reticular formation (where stimulation is neutral or aversive), or of the tegmental reticular nucleus. This system does not appear to be closely related to Olds's (1962) MFB system. In the monkey a group of self-stimulation sites is situated just dorsal to the MFB, near the medial edge of the internal capsule (Routtenberg *et al.*, 1971). In an attempt to determine whether the catecholamine-containing fibre pathways (Ungerstedt, 1971) are related to brain-stimulation reward, Crow *et al.* (1972) found that electrodes near the locus coeruleus in the pons supported self-stimulation. Electrodes near the substantia nigra and the inter-peduncular nucleus also supported self-stimulation (Crow, 1972). Unfortunately with these, as with all studies of this type, it is not possible to determine which neural structure near the electrode tip must be stimulated to produce reward. Crow *et al.* (1972) note that with the sites near the locus coeruleus the possibility that a fibre pathway associated with the mesencephalic root of the trigeminal nerve is involved in the self-stimulation cannot be excluded. Interesting reports of self-stimulation of the olfactory bulb (Phillips and Mogenson, 1969; Phillips, 1970) have appeared. Much of the neo-cortex, most of the thalamus, the reticular formation, and the cerebellum are neutral sites (Olds, 1961; Routtenberg and Malsbury, 1969; but see also sections 3.7.3 and 3.7.5). Placements near the medial and lateral lemniscus usually yield aversion (Routtenberg and Malsbury, 1969).

Intracranial self-stimulation has been reported for some similar sites of many vertebrates other than the rat. Examples of other species showing self-stimulation are the goldfish (Boyd and Gardner, 1962), pigeon (Goodman and Brown, 1966), rabbit (Bruner, 1966), cat (MFB, Roberts, 1958; Grastayán *et al.*, 1965; caudate, Justesen *et al.*, 1963; anterior nucleus of the thalamus, Grastayán and Ángyán, 1967), dog (Stark and Boyd, 1961), dolphin (Lilly and Miller, 1962), monkey

(MFB, Brodie *et al.*, 1960; amygdala, caudate, and putamen, Brady, 1960; Brady and Conrad, 1960; thalamus, Lilly, 1960; basal tegmentum, Porter *et al.*, 1959; reticular formation, Brady, 1960), and man (MFB, amygdala, septal nuclei, and intralaminar nuclei of the thalamus, Heath and Mickle, 1960; Sem-Jacobsen and Torkildsen, 1960; Bishop *et al.*, 1963; Heath, 1964).

Summary. In many vertebrates self-stimulation occurs along the extent of the MFB at rhinencephalic or limbic sites. The MFB itself is not necessarily the focal structure and in the monkey self-stimulation is obtained dorsolateral to the MFB. MFB self-stimulation is rapid, shows very little satiation, is accompanied by hyperactivity, may not give relief from intractable pain, and occurs at sites from which eating and drinking are sometimes elicited (Margules and Olds, 1962; Hoebel, 1969). Rhinencephalic self-stimulation is slow, may show satiation, is accompanied by hypoactivity, may give relief from intractable pain, and may be associated with "rebound" eating.

3.3. EFFECTS OF LESIONS ON SELF-STIMULATION

Valenstein (1966) has discussed the problems of interpreting effects of lesions on self-stimulation : e.g. it is not clear what a difference in rate of self-stimulation after a lesion may mean as rate does not always correlate with preference. Preference may be a good measure of reward, but bar-pressing rate and running speed are both affected by arousal level (see section 3.7.4). Further differences between experiments seem to be due to two factors. Firstly, only animals with imperfect bilateral lesions of the lateral hypothalamus survive the operation without special feeding. Thus an experiment which uses only animals which survive the operation may produce results which are not informative about the role of the lateral hypothalamic region. In some cases, effects have been shown by unilateral or multi-stage bilateral lesions, or by careful post-operative care. Secondly, smaller differences between experiments may be due to lesion-testing time differences. For example, some recovery occurs after anterior MFB lesions (Boyd and Gardner, 1967).

(a) *Lesions caudal to self-stimulation sites.* Careful studies by Olds and Olds (1969) and Boyd and Gardner (1967) have shown that lateral or posterior hypothalamic self-stimulation rate is decreased by small lesions in or near the MFB both anterior (in the pre-optic area) and posterior (near the interpeduncular nucleus of the midbrain) to the stimulating electrode. A posterior lesion produces the greater decrease in rate, and in contrast to an anterior lesion there is no partial recovery of rate over the few days following the lesion. In both studies ipsilateral lesions were effective, but contralateral were not. Therefore the organization of the stimulated system is unilateral between the anterior commissure and the ventral tegmental area of Tsai. Boyd and Gardner noted that their only lesion which did reduce self-stimulation rate without a correlated body weight change was to the mamillothalamic tract. These experiments receive support from the effects of xylocaine, a local anaesthetic, injected into sites anterior and posterior to MFB self-stimulation sites (Stein, 1969), which similarly decrease self-stimulation rates. The importance of regions caudal to MFB self-stimulation sites is further shown in a study by Bergquist (1970), who found in the opossum that ipsilateral lesions posterior to, but not anterior or lateral to, electrodes which produced sexual, aggressive, motivational or searching behaviour raised the threshold for the elicitation of that behaviour.

With septal self-stimulation, Schiff (1964) showed that lesions of the ventral (or mid) tegmentum blocked self-stimulation although operant bar-pressing rates increased, showing that there was no response or arousal defect. Valenstein and Campbell (1966) failed to show an effect of lateral hypothalamic lesions on self-stimulation, but their study is not conclusive because of the use of large one-stage bilateral lesions.

(b) *Of amygdala.* Kant (1969) increased septal self-stimulation rates by bilateral amygdaloid lesions. Ward (1961) performed a suction ablation of the amygdala, and after 10–20 days implanted the animals with self-stimulation electrodes in the tegmentum near the interpeduncular nucleus. The animals self-stimulated, but no measure of rate change was, of course, possible.

(c) *Of septum.* Large bilateral lesions of the septum increased the rate of self-timed intracranial lateral hypothalamic stimulation, but decreased the total stimulation time (Lorens, 1966). The voltage–current threshold for lateral hypothalamic self-stimulation was decreased by septal lesions (Keesey and Powley, 1968). Ward (1960) found no effect on rate of basal tegmental self-stimulation when four rats were tested 5–10 days after a septal suction-ablation.

(d) *Of hippocampus and fornix.* Lesions of the fornix have not affected lateral hypothalamic (Boyd and Gardner, 1967) or basal tegmental (Ward, 1960) self-stimulation. Asdourian *et al.* (1966) found that lesions in the hippocampus increased septal self-stimulation rate, and Jackson (1968) found that lesions of the ventral hippocampus lower the current levels required to maintain rates of hypothalamic self-stimulation.

(e) *Of cingulate area.* Coons and Fonberg (1963) reported that lateral hypothalamic lesions blocked cingulate self-stimulation but not vice versa.

Conclusions. Self-stimulation of, or of sites near, the MFB is decreased more by lesions posterior to than lesions anterior to the self-stimulation electrode. The effective lesions are in or near the MFB. A decreased rate or an increased threshold for self-stimulation is produced by the lesions. In contrast, septal lesions decrease the threshold or increase the rate of MFB self-stimulation. Hippocampal lesions increase the rate of septal and hypothalamic self-stimulation. Amygdaloid lesions increase septal self-stimulation rate, and do not block tegmental self-stimulation. These experiments suggest that regions caudal to self-stimulation sites may be critical in controlling whether or not self-stimulation occurs; rhinencephalic lesions appear to affect self-stimulation, producing an increased rate of responding. However, the latter conclusion probably requires revision, as anaesthetization of the amygdala and prefrontal cortex does attenuate self-stimulation (sections 3.7.2 and 3.7.3). The forebrain areas may thus modulate self-stimulation, but are probably not essential for self-stimulation, which can occur even when large areas of the forebrain, including the frontal cortex, are removed (Huston and Borbely, 1973).

3.4. THE ROLE OF PHYSIOLOGICAL CONCOMITANTS OF BRAIN-STIMULATION REWARD

Cardiovascular changes produced by brain-stimulation reward have been described for the rat by Malmo (1961), Meyers *et al.* (1963), and Perez-Cruet *et al.* (1963). Ward and Hester (1969) found that MFB self-stimulation in cats was unimpaired by bilateral surgical removal of the sympathetic chain and bilateral sectioning of the vagus and pelvic splanchnic nerves. Perez-Cruet *et al.* (1965) showed that lateral hypothalamic self-stimulation increased heart rate and blood pressure in dogs. The injection of dibenzyline, an adrenergic blocking agent, eliminated the cardiovascular effects without affecting self-stimulation. The last two experiments are evidence that cardiovascular changes are not of causal importance in self-stimulation. Further evidence against the importance of cardiovascular, endocrine or other effects which act slowly (in periods of greater than 1 s) is that self-stimulation is very sensitive to the temporal relationship of the instrumental act and the delivered stimulation, e.g. signalled reinforcement is preferred over non-signalled (Cantor and LoLordo, 1970), and extinction of self-stimulation is dependent on the time of arrival of brain stimulation relative to a bar press (Gibson *et al.*, 1965).

EEG seizure activity may appear during rhinencephalic self-stimulation (monkey, Porter *et al.*, 1959), but is very rare during posterior hypothalamic or mesencephalic self-stimulation (rat, Bogacz *et al.*, 1965). Reid *et al.* (1964) (rat) reduced seizure activity and increased self-stimulation rate with an anticonvulsant drug. This dissociation is evidence that seizure activity is not of causal importance in self-stimulation.

3.5. FIBRE DEGENERATION STUDIES

By making small lesions at self-stimulation sites and tracing the resulting orthograde fibre degeneration it is possible to draw inferences about the nature of neural systems involved in brain-stimulation reward. Routtenberg (1971) demonstrated that degeneration from rodent prefrontal cortex reward sites occupies the most medial edge of the internal capsule in a course towards the midbrain. It is in this area in the monkey, about 1 mm dorsolateral to the MFB, that

Routtenberg *et al.* (1971) obtained good self-stimulation. It may therefore be that fibres coursing caudally from the frontal cortex are involved in brain-stimulation reward. Some further evidence for this relation comes from the observation that degeneration from the monkey self-stimulation sites described above runs to the mediodorsal nucleus of the thalamus, which is connected with the prefrontal cortex, and also probably runs directly to the caudal orbitofrontal cortex (Routtenberg *et al.*, 1971). The problem with this type of study is that there is no assurance that the degenerating fibres actually fired during the self-stimulation. Which neurones fire during self-stimulation depends on the magnitude of the stimulation current, the nature of the spread of the current, and on the properties of the neurones themselves. The lesion at the self-stimulation site may destroy neurones which are not functionally related to the self-stimulation, and no check is possible. A further difficulty is that with the techniques used only orthograde degeneration has been traced, so that inferences may only be drawn about where fibres course to, not where they come from. Further, no evidence on trans-synaptic effects, i.e. on the further connections of reward pathways, can be gained with degeneration techniques. Nevertheless, experiments performed with the technique are useful in providing an indication about reward pathways. The main indication to have come so far from this work is that fibres near the MFB related to the mediodorsal nucleus of the thalamus and the prefrontal cortex are involved in brain-stimulation reward (Routtenberg *et al.*, 1971). This type of work also led to the claims that the brachium conjunctivum is a reward pathway (Routtenberg and Malsbury, 1969), and that there is a close relation between the extrapyramidal system and reward pathways (Huang and Routtenberg, 1971).

3.6. EEG DURING SELF-STIMULATION

Grastyán *et al.* (1965), working with cats, have found that hippocampal theta activity occurs during stimulation of hypothalamic reward sites. Theta EEG frequencies in the cat are 6–8 Hz. They suggest a close relationship between theta and approach-reward, and hippocampal desynchronization and withdrawal-aversion. Grastayán and Ángyán (1967), investigating self-stimulation of the anterior nucleus of the thalamus in cats, found that hippocampal desynchronization occurring during stimulation was followed by a rebound theta activity, the theta again being involved in approach. Working with immobilized rats, Routtenberg (1970) came to a similar conclusion —rewarding stimulation of brain-stem and MFB sites produced hippocampal synchronization, while rewarding subcortical telencephalic stimulation produced hippocampal desynchronization. Both effects were followed by rebound after-effects. He also found that aversive stimulation produce theta but that the theta frequency was higher than with rewarding stimulation. These studies suggest that the hippocampus may be involved in intracranial self-stimulation. Recently, Ball and Gray (1971) have found that rats showed septal self-stimulation irrespective of theta driving or blocking. Two types of septal stimulation were used, which produced driving or blocked hippocampal theta rhythm, with current intensity held constant. This experiment shows that the theta rhythm itself is not important in brain-stimulation reward.

Ball (1967) found that during self-stimulation the magnitude of potentials evoked in the trigeminal nucleus by stimulation of the infraorbital nerve was decreased. He argued from this that sensory inhibition may play a role in self-stimulation. A possible explanation is that arousal produced by the stimulation (section 3.7.4) reduced the evoked potentials.

3.7. SINGLE-UNIT ACTIVITY DURING BRAIN-STIMULATION REWARD

A good method for analysing reward pathways is to apply electrical stimulation which produces reward to a reward site, and to record from neurones in different parts of the brain to determine how the neurones are affected by the stimulation. If a single unit is fired with a short, fixed latency by the stimulus pulses, it is probably directly excited by the stimulation. This means that its axon must pass under both the recording and stimulating electrodes. A further test, for collision, can be applied to determine whether the direct excitation is antidromic or orthodromic (see below and Fig. 8). If a single unit

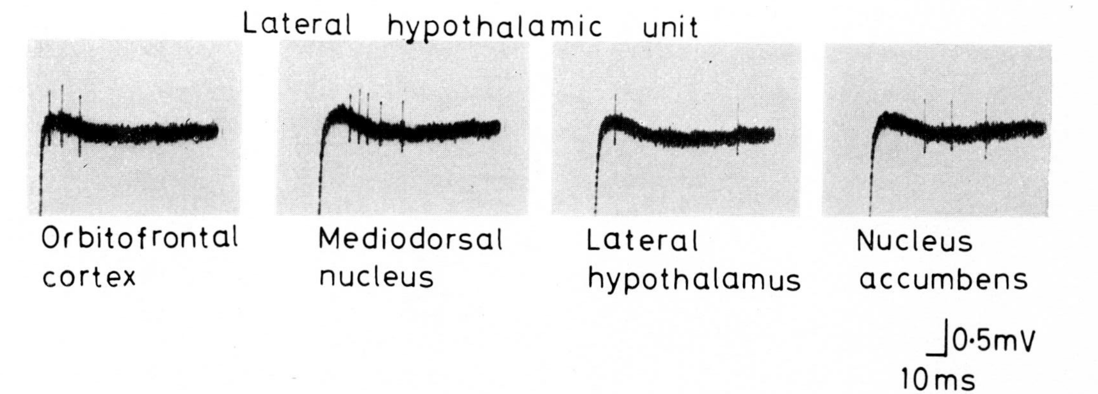

FIG. 19. This unit (vertical spikes) recorded in the lateral hypothalamus was trans-synaptically activated by rewarding stimulation of the orbitofrontal cortex (latency 6 ms), mediodorsal nucleus of the thalamus (latency 17 ms), lateral hypothalamus (latency 10 ms), and nucleus accumbens (latency 30 ms). The latency to driving from each site was relatively long and variable, and sometimes each stimulus pulse (thick vertical line at the start or left of each trace) was followed by several action potentials. Two to three superimposed traces are shown.

is fired with a longer, variable latency by the stimulus pulses, and collision cannot be demonstrated, the neurone must be trans-synaptically activated by the stimulation (Fig. 19). If the firing rate of a neurone is affected by the stimulation, yet the action potentials are not in phase with the stimulus pulses, then polysynaptic activation is likely. Using these methods, it is possible to show which neural pathways are activated by the rewarding stimulation, and to trace the effects of the stimulation across synapses through the central nervous system. Provided that the stimulating current is kept at or below the value which was sufficient for self-stimulation, it can be concluded that activated neurones were activated in self-stimulation. The specificity of the activation with respect to reward can be assessed by using animals in which the implanted electrodes do not support self-stimulation. Also, it may be stressed that in most areas of the brain single units are not activated by brain-stimulation reward. After neural systems activated in self-stimulation have been traced, further tests can be performed to determine their role in brain-stimulation reward and in naturally rewarded behaviour (see below and, for more detail, Rolls, 1974).

3.7.1. *The hypothalamic region*

Units in the hypothalamus fire during brain-stimulation reward in the squirrel monkey. This was shown in the experiments by Rolls, Burton, and Shaw described in Chapter 2. An example of one of these units is shown in Fig. 19. The superimposed traces show that the latency of firing of the unit to electrical stimulation was long and variable. The variability is probably due to variability in the delay associated with synaptic transmission, and thus indicates that the unit is trans-synaptically driven. This particular unit was activated from self-stimulation sites in the orbitofrontal cortex, mediodorsal nucleus of the thalamus, lateral hypothalamus, and nucleus accumbens. Hypothalamic units in the monkey have so far been shown to be activated in self-stimulation of these sites and of the amygdala.

In the same experiments (see Chapter 2) it was shown that some of these units are activated by natural rewards, e.g. water for a thirsty animal. Many of these units responded to visual stimuli, and fired whenever a food object was shown. These units could be involved in the orientation to food and food-seeking behaviour of the monkey. It was thus of great interest that these units were fired by brain-stimulation reward which the monkeys also worked to obtain. Other hypothalamic units activated by brain-stimulation reward responded to auditory or somatosensory stimuli, and in many other cases it was not possible to determine what affected the firing of a unit.

In the rat, Ito and Olds (1971) found units in the septal region and nucleus accumbens which were activated in posterior hypothalamic self-stimulation, and Ito (1972) recorded from hypothalamic units which were inhibited during lateral hypothalamic self-stimulation. Rolls (1970b, 1974) found that many units in the preoptic region are trans-synaptically activated during self-stimulation of the hypothalamus and the nucleus accumbens. Thus in the rat as well as the monkey, single-unit activity is greatly affected by brain-stimulation reward.

Summary. Hypothalamic neurones are fired by brain-stimulation reward of many sites and by specific natural rewards such as the taste of water or the sight of food. Brain-stimulation reward, and at least some types of naturally rewarded behaviour, e.g. eating and drinking, are diminished or abolished by lesions in these hypothalamic areas. Self-stimulation of these areas occurs. The self-stimulation, and the response of some single hypothalamic neurones to natural rewards such as food, may be modulated by drives such as hunger. Thus hypothalamic and related neurones appear to be crucial for at least some types of rewarded behaviour.

3.7.2. *Amygdala*

Recording. Single units in the basolateral region of the amygdala fire whenever a pulse is applied to self-stimulation sites in the rat (Rolls, 1972) and the squirrel monkey (experiments of Rolls, Burton, and Shaw; see Rolls, 1974). The activation may be direct, and in some cases is antidromic, i.e. with nerve impulses flowing towards the cell body. Antidromic activation can be proved by the collision test, as illustrated

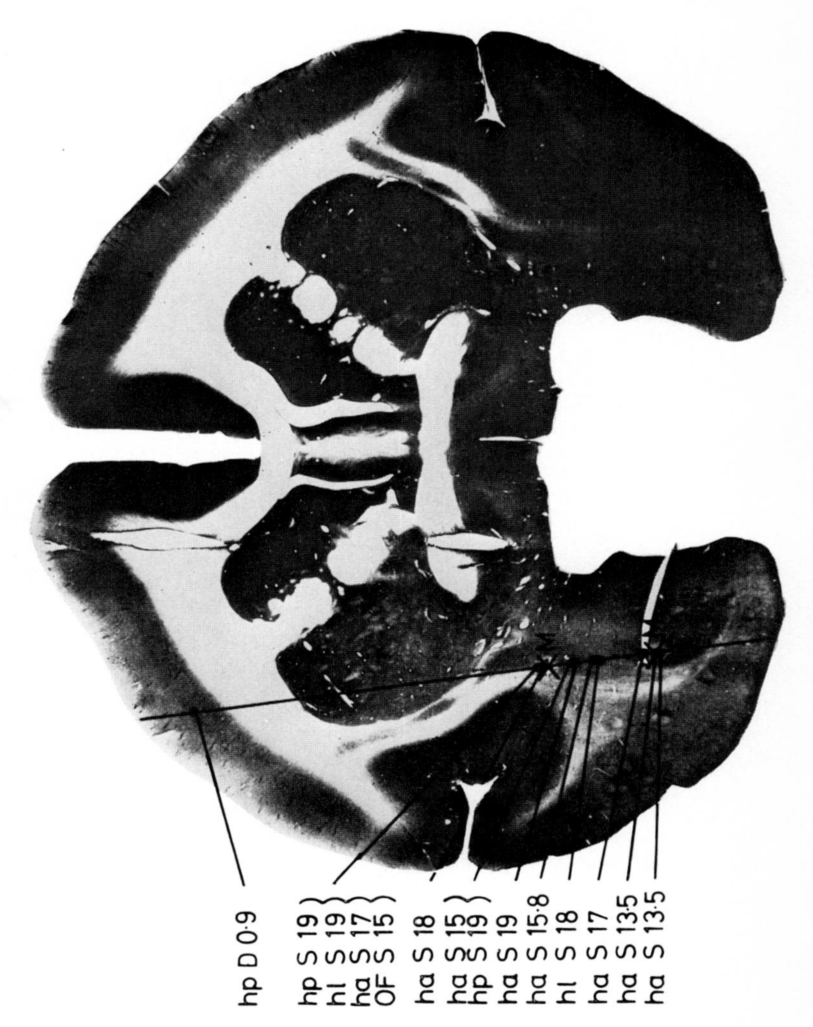

FIG. 20. A microelectrode track through the squirrel monkey lateral amygdala. Units were trans-synaptically (S) activated from reward sites in the posterior hypothalamus (hp), lateral hypothalamus (hl), anterior hypothalamus (ha), or orbito-frontal cortex (OF) with the latencies (in ms) shown. The track was located by means of the two marks (M). The track of the anterior hypothalamic self-stimulation electrode can be seen medially.

in Fig. 8. Alternatively, the activation may be trans-synaptic. An example of a microelectrode track through the amygdala in the squirrel monkey in which units activated from different hypothalamic sites were recorded is shown in Fig. 20. In recordings from 302 activated units in the rat, it was clear that the activation occurred at least from hypothalamic sites at which eating, drinking, or brain-stimulation reward was produced by the electrical stimulation, and from reward sites in the nucleus accumbens. In the squirrel monkey, activation was demonstrated from reward sites in the orbitofrontal cortex, nucleus accumbens, lateral hypothalamus, dorsomedial nucleus of the thalamus, and brain stem near the locus coeruleus. These experiments suggest that amygdaloid neurones are involved in brain-stimulation reward, and in stimulus-bound eating and drinking.

Refractory period measurements. To obtain a further indication of whether the activated amygdaloid neurones are involved in the eating, drinking, and reward (as opposed to locomotor activity, for example) elicited by the stimulation, absolute refractory period measurements were made (Rolls, 1971d, 1973). The absolute refractory period of a neurone is the shortest time interval after the excitation of one action potential when a second action potential can be excited (see, for example, Fig. 8). As axons with short refractory periods have large diameters, the refractory period can be used to characterize a population of axons of given diameters. The refractory period can thus be used to label a population of neurones. Many of the amygdaloid units directly excited by the hypothalamic stimulation had absolute refractory periods of 0.6 ms (Fig. 21). If the eating or drinking could be shown to be elicited through neurones with this refractory period, then it would increase the likelihood (but, of course, not prove) that the amygdaloid neurones mediated the eating and drinking.

To measure the absolute refractory period of the directly excited neurones which mediate eating and drinking, pulse pairs occurring repeatedly were delivered to the lateral hypothalamus while the rat was in a cage containing food or water. Only the time interval between each member of the pulse pairs (the intrapair or conditioning-test interval) was varied. If the intrapair interval (IPI) were less than the absolute refractory period, then the directly excited neurones

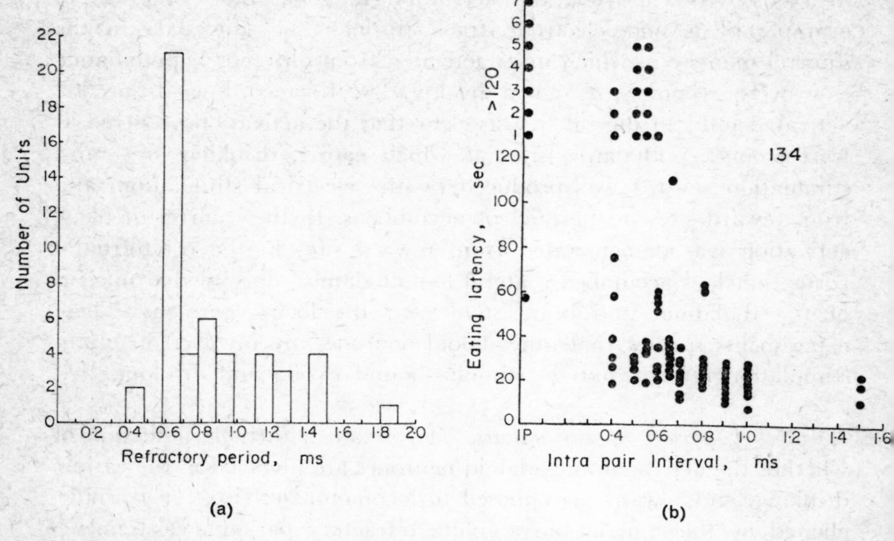

FIG. 21. (a) The absolute refractory periods of units in the region of the basolateral amygdala and pyriform cortex which were directly excited by rewarding lateral hypothalamic stimulation in the rat. (From E. T. Rolls, Activation of amygdaloid neurones in reward, eating and drinking elicited by electrical stimulation of the brain, *Brain Res.* **45,** 365–81, 1972. Reproduced with permission.) (b) Absolute refractory period of neurones involved in eating elicited by lateral hypothalamic stimulation in one rat. The latency to the onset of eating after the start of the stimulation (ordinate) is short with recurring pulse pairs in which the members of each pair are separated (abscissa) by more than 0.6 ms. (Each point represents one latency measurement.) Each time the rat failed to eat in 120 s is shown as a point at the top of the ordinate. The one pulse (1P) condition is with the second member of each pulse pair omitted. (From E. T. Rolls, Refractory periods of neurons involved in stimulus-bound eating and drinking in the rat, *J. Comp. Physiol. Psychol.* **82,** 15–22, 1973. Copyright by the American Psychological Association, and reproduced by permission.)

should fire once for every pulse pair, only little excitation would pass along the amygdaloid neurones, and little eating or drinking should occur. If the IPI is greater than the absolute refractory period, then the directly excited neurones would fire twice for every pulse pair, strong activation would reach the amygdala, and much eating or drinking would result. The effect of varying the IPI of the stimulating pulse pairs on stimulus-bound eating is shown in Fig. 21. A short latency from the start of the hypothalamic stimulation to the onset of eating represents strongly induced eating. It is clear that as IPI is increased beyond about 0.6 ms, eating starts more rapidly than at the shorter IPIs. (For statistical treatment of results and more details, see Rolls, 1973, 1974.) Thus in this animal, and in a total of five tested for eating and four for drinking, the absolute refractory period of the directly excited neurones which mediate the eating or drinking induced by hypothalamic stimulation appeared to be 0.6 ms. This result is consistent with the view that the activated amygdaloid neurones are involved in the eating and drinking elicited by lateral hypothalamic stimulation. The result does not, of course, exclude the view that other directly excited neurones are involved in the drinking, and that their involvement might be shown under different experimental conditions. The significance of the result is that it suggested that directly excited neurones with refractory periods of 0.6 ms are involved in the eating and drinking, and therefore led to further experiments on the role of the amygdala in eating and drinking.

Stimulation. Because these results suggested that the amygdala is involved in reward, electrodes were implanted in the basolateral amygdala in the squirrel monkey to test whether self-stimulation occurred. Self-stimulation did occur (see Rolls, 1974). This provided a further indication that this area is involved in brain-stimulation reward. Brain-stimulation reward has also been obtained by amygdaloid stimulation in man (Delgado, 1969) and in the rhesus monkey (Brady, 1960; Brady and Conrad, 1960), and in the medial part of the amygdala in the rat (Wurtz and Olds, 1963).

Lesions and anaesthetization. Because of these findings, Kelly (1974) made bilateral lesions in the basolateral region of the amygdala in the rat. He found that lateral hypothalamic self-stimulation was attenuated or abolished for several days. After this, recovery, measured by a return of the self-stimulation threshold to the pre-lesion level, occurred over several days. This suggests that the amygdala may modulate, but is not necessary for, brain-stimulation reward, and is consistent with the finding of Ward (1961) that bilateral ablation of the amygdala did not prevent the appearance of tegmental self-stimulation on electrodes implanted later.

To investigate the modulatory function of the amygdala further, small volumes (1–2 μl) of a local anaesthetic (5% procaine hydrochloride) were injected bilaterally into the amygdala through previously implanted cannulae (Kelly, 1974). This produces a reversible depression of brain activity for 1–2 mm round the injection site which lasts for 10–15 min (see Rolls and Cooper, 1974a). The type of finding, which occurred many times in different rats, is illustrated in Fig. 22. Within 60 s of the injection the rat stopped lateral hypothalamic self-stimulation and would not restart even when priming (free) stimulation was given (*P*). Only when a higher current was given would the rat start self-stimulating. The effect wore off after 10–15 min. Stimulus-bound eating and drinking produced by lateral hypothalamic stimulation were also attenuated or abolished by anaesthetization of the amygdala. In control experiments it was shown

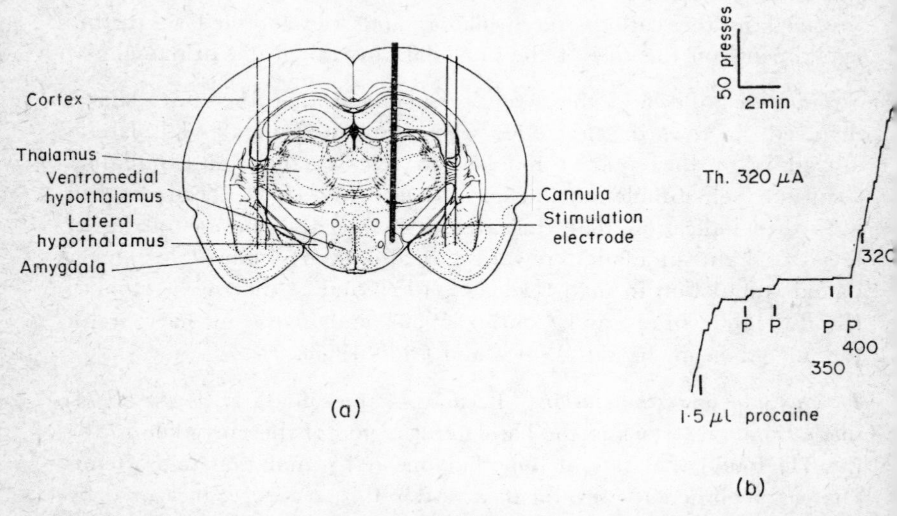

Fig. 22. (a) Local anaesthetic was applied bilaterally through the cannulae to the basolateral amygdala during reward, eating, and drinking produced by lateral hypothalamic stimulation. (b) Effect of bilateral injection of procaine HCl into the basolateral amygdala on lateral hypothalamic self-stimulation. Each time the rat presses the bar, the trace steps up (see calibration). The threshold for self-stimulation was 320 μA. Within 2 min of the injection self-stimulation ceased. It could not be reinstated, even by priming (*P*) at 350 μA. After several minutes, as the procaine became ineffective, priming at 400 μA did start self-stimulation.

that the injections had no consistent effect on locomotor activity elicited by the lateral hypothalamic stimulation. These experiments thus provide evidence that the amygdala normally exerts a specific modulatory influence on eating, drinking, and reward elicited by lateral hypothalamic stimulation.

Effects of lesions of the amygdala on natural eating and drinking. To investigate whether the basolateral part of the amygdala also modulates natural eating and drinking, bilateral lesions were made in the basolateral region in which neurones are activated in stimulus-bound eating and drinking (Rolls, B. J. and Rolls, E. T., 1973; Rolls, E. T. and Rolls, B. J., 1973). The basic controls of eating and drinking did not appear to be impaired by the lesions. Thus a cellular stimulus of drinking, hypertonic saline, which acts by withdrawing water and thus cause shrinkage of sensor cells which may be located in the preoptic region (Blass and Epstein, 1971), produced normal drinking. Drinking can also be induced by a depletion of extracellular fluid volume, by haemorrhage for example. Drinking to isoprenaline, which appears to act through the renin-angiotensin extracellular thirst mechanism (Houpt and Epstein, 1971), was also normal in the amygdala-lesioned rats, and overnight water intake was also very little changed. Similarly, the lesioned rats continued to put on body weight normally. Thus the basolateral part of the amygdala did not appear to be essential for the basic homeostatic controls of food and water intake.

Only in particular situations in which previous experience appeared to control food and water intake were abnormalities seen in the amygdala-lesioned rats. In the first situation, food preference was measured. Normal rats ate mostly familiar food, laboratory chow, and did not eat much new food, i.e. they were neophobic (Fig. 23). In contrast, rats subsequently found to have well-placed lesions in the basolateral amygdala (BL group) ate some of the unfamiliar foods (Fig. 23). Experience, and not palatability, was affected in that with repeated testing the normal rats' food preferences became similar to those of the amygdala-lesioned rats (see Rolls, E. T. and Rolls, B. J., 1973). Comparably, the amygdala-lesioned rats did not learn normally to avoid a solution of lithium chloride which produced mild sickness

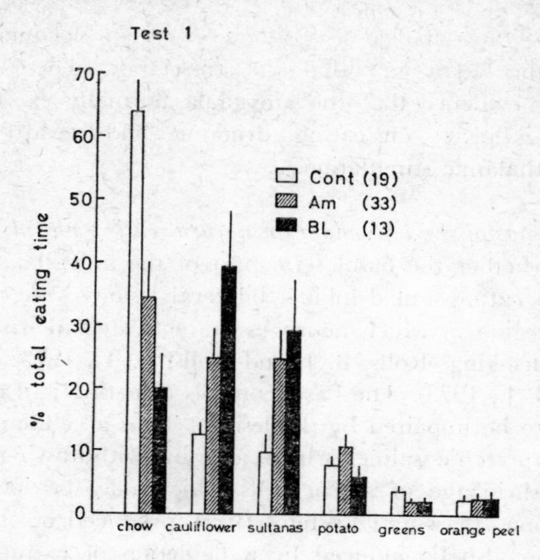

FIG. 23. Relative amounts of different foods eaten by normal (Cont) and amygdala-lesioned (Am and BL) rats. (The BL group had lesions placed particularly accurately and symmetrically in the basolateral region of the amygdala.) Food preference is indicated by the mean percentage of the total times (histograms ± SE) that the rats ate each food. The lesioned rats ate relatively less of the familiar laboratory chow than the rats with amygdala lesions. (From E. T. Rolls and B. J. Rolls, Altered food preferences after lesions in the basolateral region of the amygdala in the rat, *J. Comp. Physiol. Psychol.* **83,** 248–59, 1973. Copyright by the American Psychological Association, and reproduced by permission.)

after injection. Thus in this learned aversion situation, previous experience (feeling sick after ingesting a particular fluid) did not affect subsequent intake normally. A function of the amygdala in eating and drinking appears to be to regulate ingestion on the basis of previous experience.

Role of the amygdala in reinforcement. The experiments described above indicate that when previous experience affects ingestive behaviour, as in neophobia and learned aversion, the amygdala is involved. The amygdala receives sensory inputs, e.g. an olfactory input from the pyriform cortex in the rat (Powell *et al.*, 1965), a major visual input via the inferotemporal cortex in the monkey (Jones and Powell, 1970), and an auditory input in the cat (O'Keefe and Bouma, 1969). Thus sensory stimuli reach the amygdala–pyriform cortex region, and from here are able to influence the controls for reward in or near the hypothalamus by the amygdalo-hypothalamic pathways. It is probably in this way that the sight or smell of a new food or of a poisoned food is able to inhibit ingestion. Because of the close relation of the amygdala to reward (see above) it is likely that it controls food intake on the basis of previous experience by adjusting reward level. This would be consistent with the experiments which show that it has a modulatory but not essential role in brain-stimulation reward. Thus the amygdala can be considered as a system which allows a sensory stimulus to influence reward level. In this sense the amygdala allows stimulus–reinforcement associations to act on ingestive behaviour. The way in which the amygdala is involved in the original formation of these connections (a memory operation) is unclear, but single units in the amygdala do alter their responsiveness during the formation of stimulus–reinforcement associations. For example, some single units in the cat amygdala which had habituated to a click began to fire again consistently to the click after it had been paired with electric shock to a paw (negative reinforcement) (Ben Ari and le Gal la Salle, 1972). Similarly, in monkeys in which one signal indicated a food reward and another signal indicated electric shock, many units in the amygdala were found which responded differentially to the two signals (Fuster and Uyeda, 1971). In extinction, when the signals were no longer associated with the positive or negative reinforcement, the number of units in the amygdala which responded to the signals decreased. Thus units in the amygdala appear to be involved in one type of learning, in the formation of stimulus–reinforcement associations.

The amygdala may be involved more generally in the operation of stimulus–reinforcement associations, and a number of the features of the Klüver–Bucy syndrome in monkeys (Klüver and Bucy, 1939) may be due to a failure to make stimulus–reinforcement associations (Weiskrantz, 1956; Rolls, 1970b; Jones and Mishkin, 1972). Thus after temporal pole/amygdala lesions, monkeys became tame (the sight of a human is no longer aversive); they mouth and sometimes eat raw meat, faeces, and inedible objects (the sight and smell of the objects is not associated with edibility, and is similar to the deficit of amygdala-lesioned rats which do not show neophobia and learned aversion); and they show indiscriminate sexual behaviour, approaching male and female monkeys, and animals of other species (visual stimuli do not provide the correct guiding function of normal reinforcement). The nature of the impairment is shown more clearly in formal behavioural tests, in which animals with amygdala lesions fail to show normal conditioned avoidance learning and conditioned emotional responses (Weiskrantz, 1956; Kellicutt and Schwartzbaum, 1963; Bagshaw and Coppock, 1968). In these animals the stimulus probably does not produce the appropriate emotional response—fear. For example, in the conditioned emotional response test a tone previously paired with a shock does not interfere with (nor suppress) bar-pressing for food, i.e. conditioned fear does not disrupt the behaviour. Similarly, the amygdala appears to be involved in the operation of connections between stimuli and positive reinforcement. For example, monkeys with temporal pole/amygdala lesions (Fig. 24) do not acquire an object discrimination normally, i.e. they cannot associate easily one of two objects with food reward (Jones and Mishkin, 1972). As the amygdala does appear to be involved in the operation of stimulus–reinforcement operations in general, it is probably very important in emotional behaviour. For example, when a particular object arouses the emotion of fear, the amygdala is probably involved.

As a result of recent research, it is possible to link perceptual systems with motivational and emotional systems. Thus anatomically it is clear that there is a series of connections through the visual cortex, from area 17 (striate) and from there through several prestriate stages to the inferotemporal cortex, and from there to the amygdala (Jones and Powell, 1970). Connections from the amygdala to the hypothalamus are well known (see Eleftheriou, 1972). Behavioural experiments indicate that the prestriate cortex is involved in visual pattern discriminations, and that the inferotemporal cortex if lesioned leads to a deficit on concurrent (simultaneously learned) visual discriminations, which could indicate a memory function (Cowey and Gross, 1970; see also Mishkin, 1966, 1970; Weiskrantz, 1972; Gross, 1973). The amygdala appears to be involved in stimulus-reinforcement

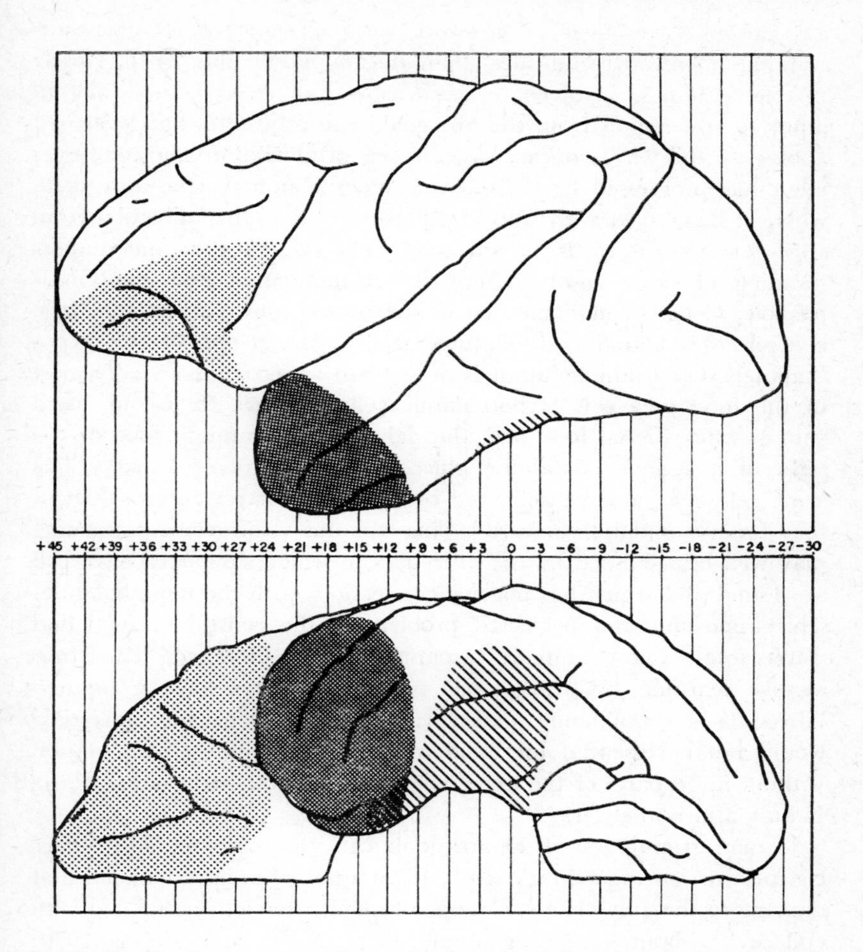

FIG. 24. In the rhesus monkey damage to the amygdala and temporal pole (dark shading) impairs the formation of stimulus–reinforcement associations, damage to the orbitofrontal cortex (light shading) impairs the disconnection of stimulus–reinforcement associations, and damage to the hippocampus and fusiform–hippocampal gyrus (cross-hatched) leads to spatial perseverations (see text). (From B. Jones and M. Mishkin, Limbic lesions and the problem of stimulus–reinforcement associations, *Expl. Neurol.* **36,** 362–77, 1972. Copyright by the Academic Press, and reproduced by permission.)

associations (see above). The work on the neural basis of brain-stimulation reward indicates that the hypothalamus or a closely related system is involved in basic aspects of reward, and that an input to this region from the amygdala can adjust reward level (see above). In this way, work on basic aspects of motivation and reinforcement has progressed back from the hypothalamus to the amygdala, where it has linked with work progressing in the other direction from sensory systems towards the amygdala. The result is that some understanding of how sensory stimuli affect motivation and emotion is possible. Coded stimuli may be linked by the amygdala to the basic controls of emotion and reinforcement in the region of the hypothalamus. For example, animals must learn to recognize visually most of the food they eat. Hypothalamic cells do learn to fire to visual stimuli which signal food (e.g. the sight of an unfamiliar food or the sight of a syringe containing glucose) (personal observations). (The same cells may also be activated by—and this may well be sufficient for—brain-stimulation reward.) This stimulus–reinforcement learning may well be mediated by the amygdala in which similar types of cell are found and which has massive connections with the hypothalamus. This important and neglected problem in the control of food and water intake, how animals recognize edible substances, may thus involve learning by amygdaloid neurones in this sensory input—amygdala—hypothalamic pathway. In this system the amygdala would not be essential for basic reinforcement, and self-stimulation without large parts of the amygdala has been obtained (Ward, 1961; Huston and Borbely, 1973).

In man, stimulation of the amygdala can lead to emotional feelings, e.g. pleasure or rage (Mark *et al.*, 1972), and it has been claimed that damage to the amygdala can reduce violent behaviour associated with epilepsy without impairing intelligence (Mark and Ervin, 1970; Narabayashi, 1972). The apparent similarity of the human and animal data are of interest, but much more research is needed before a conclusive picture can emerge even in animals (see also Valenstein, 1974).

Summary. Neurones in the basolateral amygdala are activated in self-stimulation of a number of sites in the rat and monkey as well as in eating and drinking induced by hypothalamic stimulation. Stimulation in the rat and monkey can produce reward or aversion, and in man pleasure or rage. The amygdaloid neurones directly excited by the hypothalamic stimulation have the same absolute refractory periods as the neurones through which the eating and drinking are elicited. Anaesthetization of the amygdala attenuates or abolishes the elicited eating, drinking, and reward without affecting some other types of elicited behaviour. Lesions of the amygdala attenuate these types of behaviour for only a few days, so that the amygdala appears to exert a modulatory influence on eating, drinking, and reward, and is not essential for them. Rats with lesions in the basolateral amygdala do not show normal neophobia or learned aversion. Thus a modulation on the basis of previous experience of the reward value of substances ingested may be exerted by the amygdala. The amygdala may be more generally involved in the formation of stimulus–reinforcement associations in that lesioned animals do not learn normally that a sensory stimulus signals food or electric shock. Tameness produced by amygdala lesions in animals may arise because a stimulus which normally induces fear no longer does so. That the amygdala is involved in this type of learning indicates that it plays a crucial role in emotion.

3.7.3. *Prefrontal cortex*

Neurones activated in self-stimulation have not been found in most neocortical areas. But in the prefrontal cortex, neurones activated in brain-stimulation reward are found. The prefrontal cortex in the rat was identified by Leonard (1969) by tracing anterograde degeneration from its projection nucleus, the dorsomedial nucleus of the thalamus (MD). The sulcal prefrontal cortex forms the dorsal bank of the rhinal sulcus (Fig. 25; it corresponds with the area labelled frontobasalis in Fig. 18B). It is reciprocally connected with the medial division of MD and in this respect is similar to the caudal orbitofrontal cortex of the rhesus monkey. The medial prefrontal cortex forms the medial wall of the hemisphere anterior and dorsal to the genu of the corpus callosum (Fig. 25). It is reciprocally connected with the lateral division of MD.

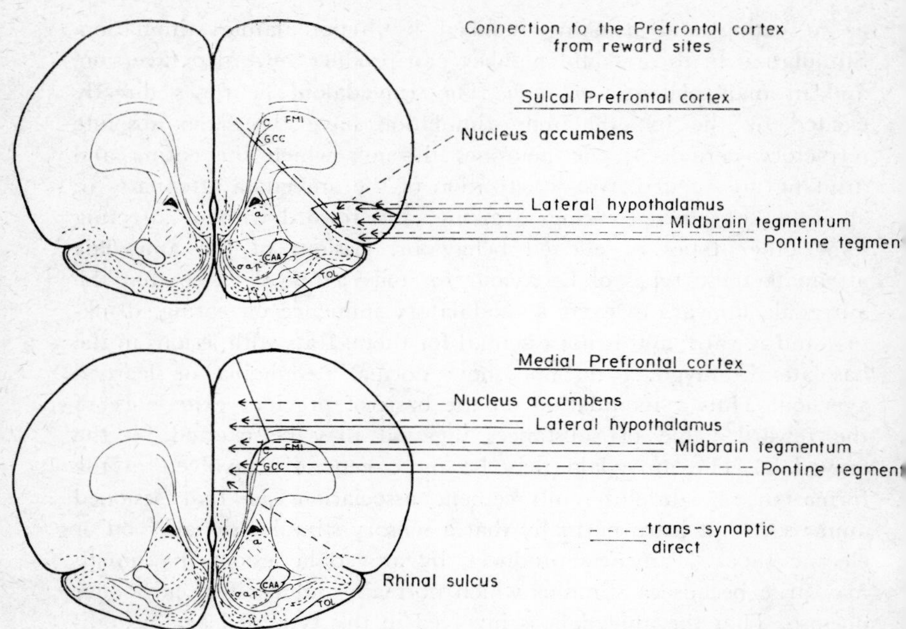

FIG. 25. Summary of connections to the prefrontal cortex from reward sites in the rat. Recordings from single units in the prefrontal cortex showed that they were directly (solid line) or trans-synaptically (dashed line) activated from the different reward sites. The approximate extents of the medial and sulcal prefrontal cortices are indicated.

Recording. Rats were implanted with electrodes and tested for self-stimulation. Then recordings from single units in the prefrontal cortex were made in the anaesthetized animal while stimulation was applied to the self-stimulation electrodes. To ensure that all the neurones activated in the electrophysiological experiment had also been activated during self-stimulation, only currents near the threshold value for self-stimulation were used during the electrophysiology. To determine whether activation of neurones in the prefrontal cortex was characteristic of self-stimulation, a number of different self-stimulation sites were used, and also effects from a non-reward site were investigated. In previous electrophysiology performed on the prefrontal cortex during brain-stimulation reward, Ito and Olds (1971) had

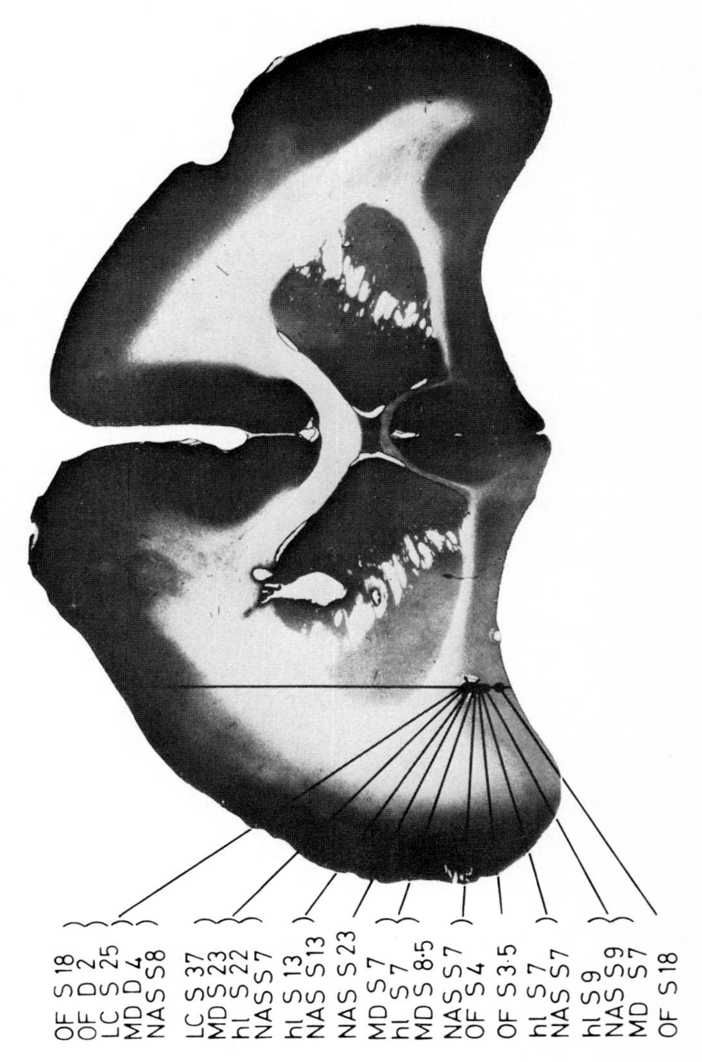

FIG. 26. Microelectrode track through the orbitofrontal cortex of the squirrel monkey. Single units were directly (D) or trans-synaptically (S) activated with the latencies shown (in ms) from self-stimulation sites in the orbitofrontal cortex (OF), nucleus accumbens (NAS), lateral hypothalamus (hl), mediodorsal nucleus of the thalamus (MD), or in the region of the locus coeruleus (LC). A lesion marked the position of the track.

found approximately six neurones in the pregenual frontal (four probably in the medial prefrontal) cortex which were activated in hypothalamic self-stimulation.

An example of a recording of a single unit in the sulcal prefrontal cortex is shown in Fig. 8. The unit was directly excited by stimulus pulses applied to a self-stimulation site near the locus coeruleus in the pontine tegmentum. The unit was classed as directly excited because it followed each stimulus pulse with a short fixed latency (of 2 ms), it had a short absolute refractory period shown by the intermittent firing to the second stimulus pulse when it followed a first pulse by 0.82 ms (Fig. 8) and because it showed collision and was therefore antidromically activated (Fig. 8). With antidromic activation action potentials initiated normally and travelling orthodromically away from the cell body collide with and abolish action potentials travelling antidromically from a site of electrical stimulation towards a cell body (Fig. 7). Thus in Fig. 8, where a spontaneous action potential precedes (is on the left of) the stimulus pulse by 2.0 ms or less, the electrically elicited action potential does not reach the recording electrode. This is a useful method for analysing the arrangement of activated neurones. In this case the neurone has a cell body in the sulcal prefrontal cortex and an axon which courses to the stimulating electrode in or near the locus coeruleus at the far end of the brain.

Many neurones in the sulcal and the medial prefrontal cortices were shown to be directly (most probably antidromically) excited in brain-stimulation reward of many different sites (see Fig. 25 and Rolls and Cooper, 1973, 1974b; Rolls, 1974). Other units were trans-synaptically activated, with longer, more variable latencies, from different reward sites (dotted connections in Fig. 25). (Further criteria for distinguishing directly from trans-synaptically activated units are given in Rolls, 1971a, 1972, 1974. An example of a trans-synaptically activated unit is shown in Fig. 19.)

In the orbitofrontal cortex of the monkey, and in the mediodorsal nucleus of the thalamus, single units are also activated by brain-stimulation reward. An example of a microelectrode track through the squirrel monkey orbitofrontal cortex is shown in Fig. 26. Units in these regions have been shown to be activated in self-stimulation of the nucleus accumbens, lateral hypothalamus, amygdala, and locus coeruleus, as well as the orbitofrontal cortex and the mediodorsal nucleus of the thalamus (experiments of Rolls, Burton, and Shaw; see Rolls, 1974). These experiments then suggest that some parts of the prefrontal cortex are involved in brain-stimulation reward.

Anaesthetization. To investigate this suggestion, the sulcal prefrontal cortex in the rat was anaesthetized bilaterally during self-stimulation. Injections of 1–2 μl of 5% procaine hydrochloride in or near the sulcal prefrontal cortex attenuated self-stimulation of the lateral hypothalamus and the pontine tegmentum (Rolls and Cooper, 1973, 1974a). The attenuation was measured by a decreased rate or cessation of self-stimulation or by an increase in the current necessary to maintain self-stimulation. Injections away from the sulcal prefrontal cortex were less effective in attenuating self-stimulation. It can be concluded that anaesthetization of the sulcal prefrontal cortex, or of a region near it, attenuates brain-stimulation reward.

Stimulation. Self-stimulation of the medial prefrontal cortex (Routtenberg, 1971; Rolls and Cooper, 1973) and of the sulcal prefrontal cortex (Routtenberg and Sloan, 1972; Rolls and Cooper, 1974a) in the rat occurs. It has also been shown that self-stimulation in the orbitofrontal cortex and of the mediodorsal nucleus of the thalamus occurs in squirrel monkeys (experiments of Rolls, Burton, and Shaw; see Rolls, 1974).

Role of the prefrontal cortex in reinforcement. The experiments described above, together with the observations that anterograde degeneration from medial prefrontal self-stimulation sites in the rat courses through self-stimulation sites near the medial forebrain bundle (Routtenberg, 1971), and that anterograde degeneration from diencephalic self-stimulation sites in the monkey courses to the mediodorsal nucleus of the thalamus and perhaps to the orbitofrontal cortex (Routtenberg *et al.*, 1971), indicate that the prefrontal cortex is involved in brain-stimulation reward. It may not be essential for brain-stimulation reward. For example, even a rat incapacitated by removal of much of its neocortex can learn to move its tail in order to obtain hypothalamic stimulation (Huston and Borbeley, 1973).

The function of the prefrontal cortex in reward can be considered together with the known functions of the prefrontal cortex. In the monkey, lesions of the orbitofrontal cortex (see Fig. 24) lead to poor performance in an object discrimination reversal task because the lesioned monkeys continue to select the previously rewarded but now unrewarded stimulus (Jones and Mishkin, 1972); to poor extinction measured by prolonged bar-pressing when reward is no longer available (Butter, 1969); and to deficits in a go/no-go task in which the lesioned monkeys cannot learn not to reach out on a no-go trial. These tasks appear to share an "unlearning" factor. In each task, the lesioned monkeys appear to be unable to change their behaviour when reward is no longer associated with a stimulus or a type of behaviour. The animals cannot break a stimulus-reinforcement association—that is, "unlearn". In order to perform this suggested function, the orbitofrontal cortex would need to know when reward arrived so that it could operate on the omission of a reward. The electrophysiological experiments show that the orbitofrontal cortex (and the sulcal prefrontal cortex in the rat) does receive a signal when reward arrives. The orbitofrontal cortex could then influence other reward neurones (e.g. in the hypothalamus) if a stimulus is no longer associated with reinforcement. Thus one reason why the orbitofrontal cortex (and perhaps also the sulcal prefrontal cortex in the rat) is closely connected with brain-stimulation reward may be that it normally signals when expected reward does not arrive.

In man, frontal lobe damage can lead to perseveration of strategies, seen in inability to change the categories into which cards are sorted, and in inability of the patients to learn a maze correctly (see Nauta, 1971; Luria, 1973). Perseveration of a strategy when inappropriate is very similar to the failure to disconnect stimulus–reinforcement associations seen in orbitofrontal monkeys. Whether this type of perseveration is related to orbitofrontal damage is unclear. An interesting characteristic of the human patients, which may have relevance to the interpretation of animal experiments, is that the patients realize and state that their strategy is wrong yet still persist in their strategy.

If the hypothesis that a function of the orbitofrontal cortex is to disconnect stimulus–reinforcement associations is correct, then the orbitofrontal cortex should be involved in emotional behaviour. Phobias or compulsive obsessional behaviour could be viewed as abnormal persistence of stimulus–reinforcement associations. Emotional reactions (e.g. frustration) to the absence of an expected reward could also be expected to be influenced by the orbitofrontal cortex. These points are very tentative, and more animal research is needed before extrapolations to human behaviour can be made (see also Valenstein, 1974). The role of the frontal lobes in human emotional behaviour is certainly unclear. During an investigation of the effects of frontal lobe damage on delayed-response tasks in chimpanzees, Jacobsen (1936) noted that one of his animals became calmer and showed less anger and frustration after the operation when reward was not given. (A delayed-response deficit follows damage to the dorsolateral frontal cortex in the monkey—Goldman *et al.*, 1971—with which we are not concerned here.) Hearing of this emotional change, Moniz argued that anxiety, irrational fears, and emotional hyperexcitability could be treated in man by damage to the frontal lobe. He operated on twenty patients and published an enthusiastic report of his findings (Moniz, 1936; Fulton, 1951). This rapidly led to the acceptance of the surgical procedure, and more than 20,000 patients were subjected to "lobotomies" or "leucotomies" of varying extent during the next 15 years. Although irrational anxiety or emotional outbursts were sometimes controlled, intellectual deficits and other side-effects were often apparent (Rylander, 1948; Valenstein, 1974). Thus these operations have been essentially discontinued and are replaced by drug treatment or behaviour therapy (see, for example, Marks, 1969).

One interpretation of the results described above is that after orbitofrontal damage the omission of an expected reinforcement produces an abnormally small emotional reaction (e.g. only mild frustration), and a failure to alter subsequent behaviour as a result of the omission. A decreased emotional response to pain may also follow frontal lobe damage. Patients who have undergone a frontal lobotomy report that after the operation they still have pain but that it does not bother them (Freeman and Watts, 1950; Melzack, 1973). Perhaps comparably, monkeys with orbitofrontal damage do not learn to normally avoid punishment, i.e. they have an avoidance deficit (Tanaka, 1973).

Thus at present it appears that perseveration produced by a failure to disconnect stimulus–reinforcement associations, together with lack of emotional response when reward is omitted or punishment is given, follows damage to the orbitofrontal cortex. This conclusion must be tentative in view of the diversity of functions represented in the frontal lobes (Teuber, 1972), and great care must be taken over conclusions which may be relevant to man (Valenstein, 1974).

Summary. Neurones in the sulcal and the medial prefrontal cortices in the rat are activated in self-stimulation of many different brain sites. Self-stimulation of the sulcal and the medial prefrontal cortices occurs. Bilateral anaesthetization of the region of the sulcal prefrontal cortex attenuates brain-stimulation reward. Therefore the sulcal prefrontal cortex (and possibly also the medial prefrontal cortex) may be involved in brain-stimulation reward. Similarly, in the monkey, neurones in the orbitofrontal cortex are activated during self-stimulation, and the orbitofrontal cortex supports self-stimulation. In the monkey, the orbitofrontal cortex appears to be involved in breaking a learned association between a stimulus and reinforcement. Thus the orbitofrontal cortex in the monkey and the sulcal prefrontal cortex in the rat may be closely related to brain-stimulation reward because these parts of the brain must normally monitor reward so that they can operate as soon as reward fails to occur. These parts of the cortex can be considered as high-level controls of reward which are concerned with changing behaviour when reward is no longer associated with the behaviour. Because of the role in reward, the monkey orbitofrontal and rat prefrontal cortex may not be essential for reward. The function of these areas in reward is analogous to that of the amygdala, except that the amygdala may be concerned with the formation (as opposed to the disconnection) of stimulus–reinforcement associations.

3.7.4. *Midbrain neurones and arousal*

Neurones with cell bodies in the midbrain and pons are activated by rewarding stimulation of the lateral hypothalamus. The stimulation also produces arousal, measured by desynchronization (absence of large waves) of the EEG, and the activation of arousal units whose firing rate is correlated with the EEG change (see Fig. 27 and Rolls, 1971a). Arousing stimuli such as a pinch to the hind leg and the inhalation of amyl nitrite also produce these effects. These midbrain units are not activated, and EEG desynchronization is not produced by rewarding stimulation of the nucleus accumbens (Rolls, 1971c) or by non-rewarding stimulation of sites near the lateral hypothalamus from which motor effects are elicited (Rolls, 1971a). These results, and evidence from tracing the effects of electrical stimulation using refractory period measurements (Rolls, 1971a), suggested that arousal is produced by rewarding lateral hypothalamic stimulation because it activates a midbrain arousal system (Rolls, 1971a, c, 1974). The

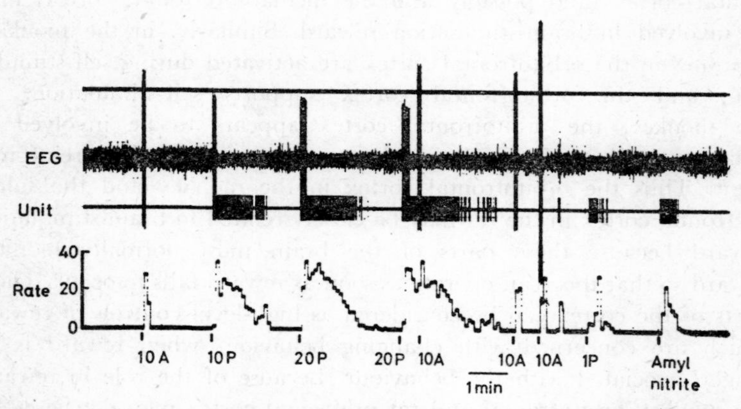

FIG. 27. Arousal is not produced by stimulation applied to a nucleus accumbens septi self-stimulation site. One, ten, or twenty trains of stimulation applied to the posterior, MFB, self-stimulation electrode (1P, 10P, 20P) increased the firing rate of the unit (see "unit" and "rate" traces) and produced cortical desynchronization. Ten trains applied to the anterior, nucleus accumbens electrode (10A) did not produce a similar arousal effect. (Only stimulus artefact was counted in the rate trace.) The same stimulation did not inhibit the arousal effect (20P followed by 10A). Each 0.3 s train of pulses at 100 pulses/s was just above the threshold of self-stimulation. The unit, which was indirectly driven by MFB stimulation, had general arousal properties in that its firing rate was correlated with cortical desynchronization, and stimuli which altered arousal level (e.g. amyl nitrite) always produced a correlated change in the firing rate of the unit. (From E. T. Rolls, Contrasting effects of hypothalamic and nucleus accumbens septi self-stimulation on brain stem single unit activity and cortical arousal, *Brain Res.* **31,** 275–85, 1971.)

arousal decays over a period of minutes following several trains of hypothalamic stimulation (Rolls, 1971a, c, 1974) and is reflected in unanaesthetized rats as stimulus-bound locomotor activity (Rolls and Kelly, 1972). The arousal produced is important in analysing self-stimulation because it affects the rate of self-stimulation (Rolls, 1971b). The influence of arousal on self-stimulation rate is of importance for example in interpreting the effects of pharmacological agents on brain-stimulation reward (see Chapter 4). This arousal system is not of primary importance in brain-stimulation reward, as it is not activated by rewarding stimulation of the nucleus accumbens, and its activation is therefore not an essential aspect of brain-stimulation reward (Rolls, 1971c, 1974).

3.7.5. *Other brain areas* (see Cooper and Rolls, 1974; Rolls, 1974)

Neurones in the hippocampus (and entorhinal cortex) are trans-synaptically activated with long latencies (e.g. 30–100 ms) by brain-stimulation reward. This activation is not necessary for but may affect brain-stimulation reward, as lesions of the hippocampus generally facilitate self-stimulation measured by the rate of bar-pressing. Neurones in the cingulate cortex are activated (some antidromically) with short latencies by rewarding hypothalamic stimulation. The hypothalamus may be necessary for cingulate self-stimulation but not vice versa. Single units in the reticular nucleus of the thalamus are activated directly or trans-synaptically by rewarding hypothalamic stimulation. Their role in self-stimulation is not known. Neurones in the lateral parts of the pons and medulla (near or in the nucleus reticularis parvocellularis and locus coeruleus) are activated directly or trans-synaptically from many different reward sites, but their role in self-stimulation is not clear. Neurones in the more medial parts of the pons and medulla (e.g. the nucleus reticularis gigantocellularis) are activated by both self-stimulation and during motor effects produced from non-reward sites, and may be related to the elicitation of motor effects.

3.8. SUMMARY (see Fig. 28)

The hypothalamus is a critical area for brain-stimulation reward in that it contains self-stimulation sites, lesions in or caudal to it can attenuate self-stimulation, fibres from many self-stimulation sites course to or close to it, and single neurones in the hypothalamus fire during self-stimulation of many different sites. Some hypothalamic neurones which are activated by specific natural rewards, e.g. water for a thirsty animal or the sight of a banana for a hungry animal, are also activated by brain-stimulation reward. It is suggested that brain stimulation can provide reward because it activates neurones of this general type—neurones fired by specific natural rewards. The effects of natural stimuli and of brain stimulation on some specific natural reward neurones can probably be gated, in that food or brain stimulation at one site may only provide reward if the animal is hungry, and water or brain stimulation at another site may only provide reward if the animal is thirsty.

The amygdala is related to brain-stimulation reward in that neurones in the amygdala are activated in brain-stimulation reward of some sites, self-stimulation of some sites can be attenuated by anaesthesia of the amygdala, and self-stimulation of the amygdala can be obtained. The amygdala may not be essential for brain-stimulation reward in that lesions of the amygdala may not permanently abolish self-stimulation. It is suggested that the amygdala is related to brain-stimulation reward because it is concerned with learning about which stimuli are rewarding (or punishing). Thus animals with bilateral damage to the amygdala have difficulty in distinguishing visually between food and non-food objects and in learning to avoid punishment or a food previously associated with sickness. Further, neurones in the amygdala may only continue to fire to environmental stimuli which are associated with reward or punishment—they may learn stimulus–reinforcement associations. The amygdala is important in emotional behaviour in that learned reactions to rewards (which are pleasant) and to punishments (which are aversive) must be crucial to normal emotional behaviour, and that stimulation in the human amygdala can produce general emotional responses. This role of the amygdala in stimulus–reinforcement learning not only suggests why it is related to brain-stimulation

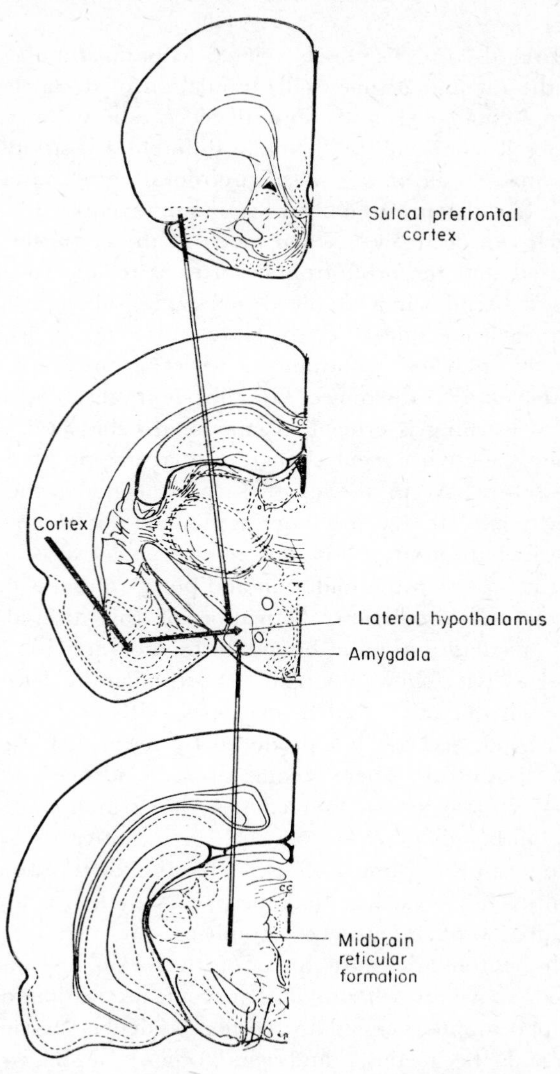

FIG. 28. Schematic diagram showing three neural systems involved in self-stimulation (see section 3.8).

reward, but also clarifies the neural basis of one very important type of learning.

The prefrontal cortex is closely related to brain-stimulation reward in that in the rat and monkey self-stimulation of it can be obtained, and electrophysiological and fibre degeneration experiments show that it is closely connected with many different self-stimulation sites. In the rat anaesthesia of the sulcal prefrontal cortex attenuates self-stimulation, but the prefrontal cortex is not essential for self-stimulation, which can occur even when much of the neocortex is ablated. It is suggested that the orbitofrontal cortex is related to brain-stimulation reward because it is involved in learning about when rewards (or punishments) no longer occur. Thus after damage to the orbitofrontal cortex, monkeys continue to select a previously rewarded stimulus—they fail to disconnect stimulus–reinforcement associations. This type of learning is crucial to emotional behaviour, which must change when the environment changes so that inappropriate strategies do not persevere. As in the case of the amygdala, the prefrontal cortex may modulate reward, but in relation not to the formation but to the disconnection of stimulus–reinforcement associations.

Through neurones in the midbrain and pons, arousal is produced by hypothalamic self-stimulation. Activation of this arousal system at least partly mediates stimulus-bound locomotor activity in that the arousal and activity show a similar post-stimulation decay, are produced through directly excited neurones with the same absolute refractory period, and are not produced by rewarding stimulation of the nucleus accumbens. There is some similar evidence that activation of this arousal system partly mediates the priming effect in self-stimulation, although other factors such as incentive motivation may also contribute to the priming effect. Activation of this arousal system is not essential for reward in that the arousal system is not activated in self-stimulation of the nucleus accumbens.

Although neurones in most areas of the brain are not activated as described above in self-stimulation, some activation of neurones in the hippocampus, cingulate cortex, reticular nucleus of the thalamus, and the medulla and caudal pons by brain-stimulation reward has so far been found. The function of activation of these neurones in self-stimulation is not yet known.

CHAPTER 4

PHARMACOLOGY OF SELF-STIMULATION

RESEARCH on the possible neurotransmitters and modulators which mediate and affect self-stimulation is very active. In this area anatomical, neurophysiological, and behavioural findings can be linked with pharmacology. The results are relevant to emotional behaviour and its disorders.

4.1. INTRODUCTION

Much of the research on the pharmacology of brain-stimulation reward has centred on the function of catecholamines in self-stimulation. The catecholamines are of particular interest in this context in that they are involved in emotional behaviour according to the catecholamine hypothesis of affective disorders (Schildkraut, 1965; Schildkraut and Kety, 1967). The two catecholamines found in the brain with which we are concerned are noradrenaline (NA) and dopamine (DA). A schematic view of a noradrenergic synapse showing some of the ways in which pharmacological agents used to analyse self-stimulation are believed to affect its function is shown in Fig. 29 (see also Moore, 1971). Tyrosine is converted to dihydroxyphenylalanine (DOPA) in the presence of the enzyme tyrosine hydroxylase. The drug α-methyl-*p*-tyrosine blocks this step, and thus inhibits the synthesis of NA (and DA). DOPA is converted to DA, and DA is converted to NA in the presence of the enzyme dopamine-β-hydroxylase. This step is inhibited by disulfiram (or by diethyldithiocarbamate), so that the administration of disulfiram prevents the synthesis of NA but not of DA. Noradrenaline is normally released from the presynaptic membrane in vesicles when an action potential

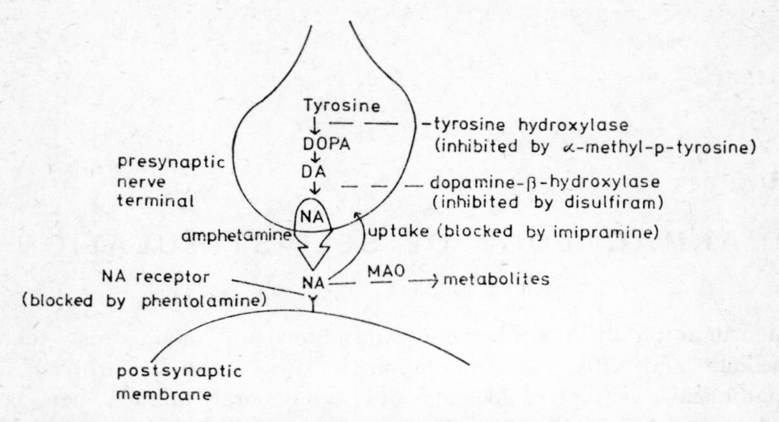

FIG. 29. Schematic diagram showing how pharmacological agents used to analyse self-stimulation are believed to affect a noradrenergic synapse (see section 4.1). DOPA, dihydroxyphenylalanine; DA, dopamine; NA, noradrenaline (or norepinephrine); MAO, monoamine oxidase. Amphetamine facilitates the release of NA.

occurs, and travels across the synaptic cleft to excite NA receptors on the post-synaptic membrane. The drug amphetamine releases NA from the presynaptic membrane. The effect of NA on the post-synaptic receptor is blocked by phentolamine, and one of the many effects of the major tranquillizers (or neuroleptics) such as chlorpromazine and haloperidol used to treat schizophrenia is NA-receptor blockade. The main process which removes NA from the synaptic cleft is uptake into presynaptic structures and this is blocked by tricyclic antidepressants such as imipramine, and also by amphetamine. Another process which removes NA from the synaptic cleft is metabolism of NA using, for example, the enzyme monoamine oxidase. Monamine oxidase (MAO) inhibitors (such as iproniazid and tranylcypromine) are also antidepressants. Thus antidepressants facilitate noradrenergic (and also dopaminergic) mechanisms.

There may be dopaminergic synapses in the brain which are schematically similar to noradrenergic synapses and in which DA is released and not NA. Many pharmacological agents have effects on both noradrenergic and dopaminergic synapses, and therefore care is required in interpreting the actions of these agents. Dopamine is formed on the same biosynthetic pathway as NA, and thus α-methyl-*p*-tyrosine depletes DA-containing neurones of DA. Amphetamine releases DA from the presynaptic structures. Spiroperidol (spiperone) and pimozide block DA receptors (and only block NA receptors if the dose is increased 1000 times—Andén *et al.*, 1970). The neuroleptic drugs such as chlorpromazine block DA and NA receptors to different extents, but some of the neuroleptics, such as pimozide, block mainly DA receptors (Andén *et al.*, 1970). The antidepressant drugs such as imipramine and the monoamine oxidase inhibitors lead to an accumulation of DA.

A number of comparisons between noradrenergic and dopaminergic mechanisms can usefully be remembered when interpreting the actions of pharmacological agents on self-stimulation (Figs. 29 and 30).

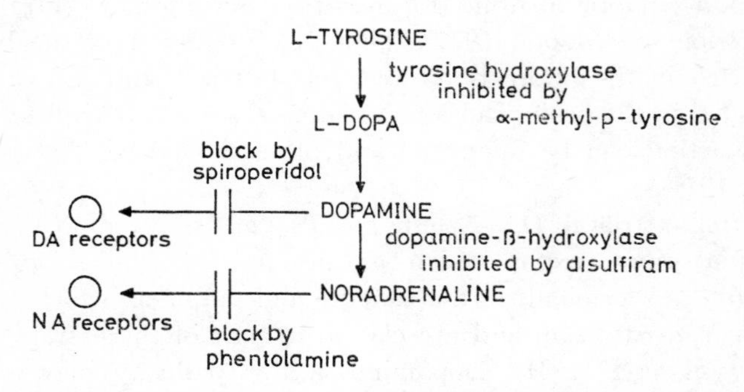

Fig. 30. Effects of some pharmacological agents on catecholamines.

α-Methyl-*p*-tyrosine inhibits the synthesis of NA and DA, while disulfiram inhibits the synthesis of NA but not DA. Amphetamine releases both NA and DA. Phentolamine blocks NA-receptors but not DA-receptors. Spiroperidol and pimozide block DA-receptors and not NA-receptors. Chlorpromazine and haloperidol block both DA- and NA-receptors to some extent (as well as having other actions). The drug 6-hydroxydopamine can cause degeneration of both NA and DA nerve terminals. The antidepressant drugs facilitate both NA and DA mechanisms.

The catecholamines NA (norepinephrine) and DA are located in brain cells whose course has been traced using histofluorescence techniques (see Ungerstedt, 1971). Ascending noradrenergic pathways of present interest in the control of self-stimulation are as follows. Noradrenaline cell bodies in the locus coeruleus (cell group A6 according to nomenclature of Dahlström and Fuxe, 1965) give rise to ascending pathways which form a dorsal bundle in the pons and midbrain, ascend through the medial forebrain bundle, and innervate the cortex and hippocampus (Fig. 31). The "ventral NA pathway" arises from cell groups A1, A2, A5, and A7, and has terminals in the lower brain stem, midbrain, and hypothalamus and preoptic area (Fig. 31). Noradrenaline in the brain satisfies a number of criteria required to establish it as a neurotransmitter, i.e. as a substance released by action potentials at synaptic terminals of one neurone which crosses the synaptic cleft to influence action potential formation in the post-synaptic neurone (for a useful description of central neurotransmission see Moore, 1971). Thus NA is present in the brain, is synthesized in the brain, is released by electrical stimulation, can be inactivated in the brain, can affect neuronal activity if applied locally, and its action can be mimicked and blocked pharmacologically (see Moore, 1971).

The nigro-striatal DA system has its cells of origin (A9) in the substantia nigra, axons which ascend through the lateral hypothalamus, and terminals in the caudate and putamen, which are parts of the corpus striatum and are classed as part of the extrapyramidal motor system (Fig. 31). Dopamine in this pathway may act as a neurotransmitter, and is probably involved in Parkinson's disease, symptoms of which may include lack of voluntary movement (Hornykiewicz, 1973). The mesolimbic DA system has its cells of origin (A10) dorsal to the nucleus interpeduncularis and medial to the substantia nigra, axons which ascend just dorsal to the medial forebrain bundle, and terminals in the nucleus accumbens, nucleus interstitialis striae terminalis, and tuberculum olfactorium (Fig. 31).

4.2. EVIDENCE THAT CATECHOLAMINES ARE INVOLVED IN SELF-STIMULATION

There is considerable evidence that catecholamines affect self-stimulation rate, much of it accumulated and reviewed by Stein (1967, 1969). Drugs which increase or mimic catecholamines increase

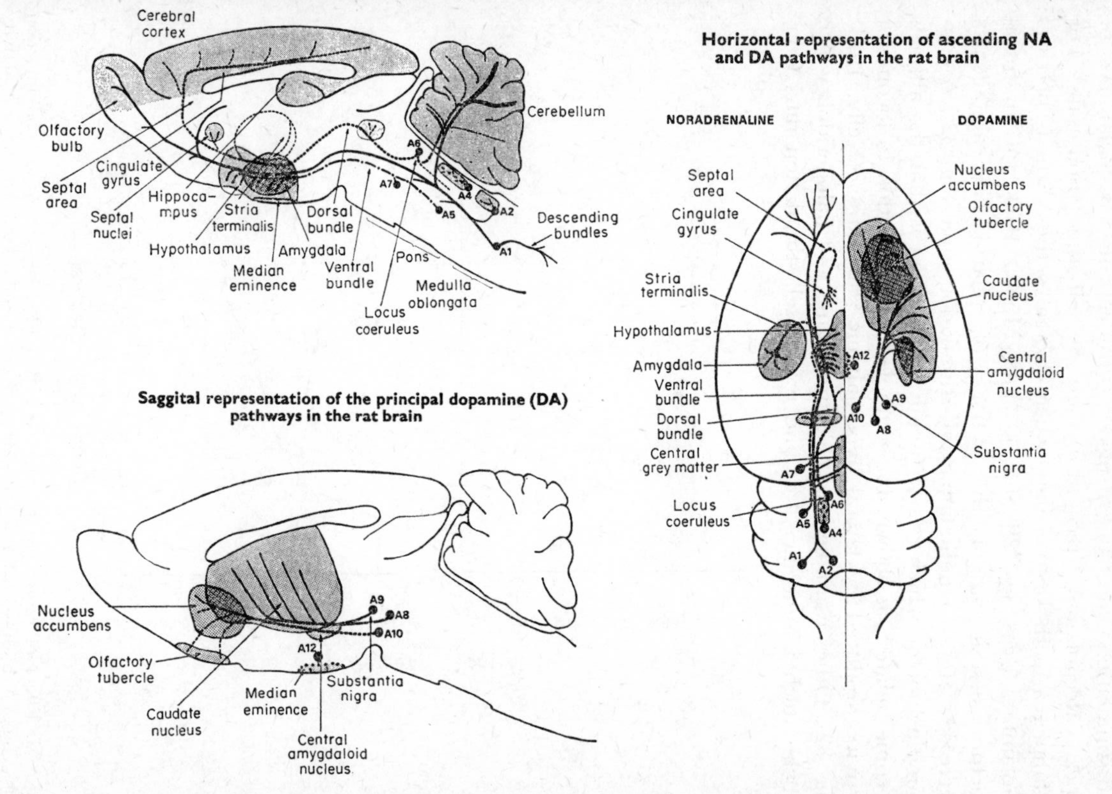

FIG. 31. Course of NA-containing and DA-containing neurones in the rat brain. (After Ungerstedt, 1971; and from B. G. Livett, *Br. Med. Bull.* **29,** no. 2, 93–99, 1973.

self-stimulation rate, and drugs which decrease catecholamines or block catecholamine receptors decrease self-stimulation rate. For example, α-methyl-*p*-tyrosine, which decreases the synthesis of catecholamines (see Figs. 29 and 30; and Moore, 1971, for an introductory description of the actions of pharmacological agents), decreases lateral hypothalamic self-stimulation rate (Fig. 32). A confounding factor which is not usually emphasized is that the animals become sedated, as shown, for example, by a reduction in locomotor activity produced by lateral hypothalamic stimulation (Kelly, 1974; Fig. 32). Other examples are chlorpromazine and haloperidol, which block catecholamine receptors and reduce self-stimulation rate. Con-

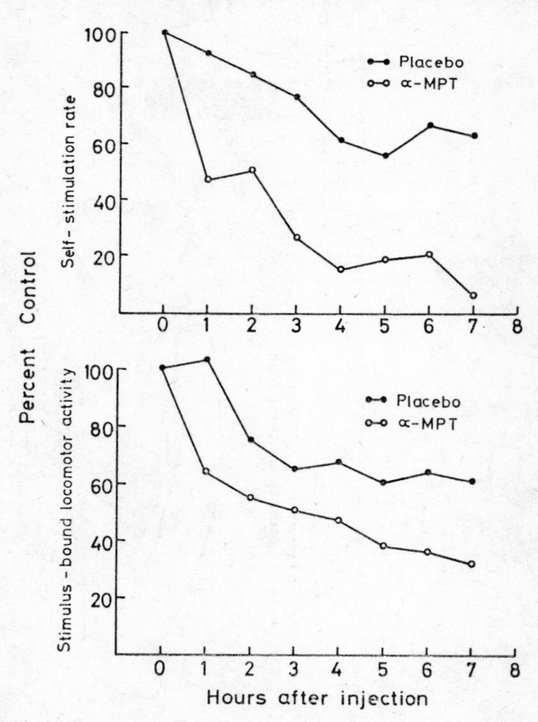

FIG. 32. Effect of α-methyl-*p*-tyrosine (α-MPT) (200 mg/kg) or control injection (placebo) on both self-stimulation rate and on stimulus-bound locomotor activity measured in the same rats. (After Kelly, 1974.)

versely, amphetamine, or tetrabenazine in combination with a monoamine oxidase inhibitor, both of which release catecholamines, increase self-stimulation rate. It is notable that amphetamine also increases arousal, measured by an increase of stimulus-bound locomotor activity (Kelly, 1974).

4.3. DOES EITHER NORADRENALINE OR DOPAMINE MEDIATE BRAIN-STIMULATION REWARD?

The above and other evidence has been interpreted as showing that the release of NA mediates brain-stimulation reward (e.g. Stein, 1969, 1971). There are two major problems with this. Firstly, the pharmacological treatments affect DA as well as NA (for review see Rolls, 1974), and therefore do not show that the relevant catecholamine is NA. Secondly, many factors other than reward affect self-stimulation rate (or the threshold of self-stimulation), so that a side-effect, e.g. sedation, could have produced a change in self-stimulation rate. To analyse this more precisely, adequate controls for side-effects must be performed, and pharmacological agents with more specific effects on particular catecholamines must be used.

There have been few studies of the effects on brain-stimulation reward of treatments which alter the activity in specific catecholaminergic systems. One agent, disulfiram, which depletes the brain of NA but not DA (see Figs. 29 and 30) by inhibiting the enzyme dopamine-β-hydroxylase (Musacchio *et al.*, 1966; Goldstein and Nakajima, 1967), can abolish self-stimulation (Wise and Stein, 1969). This indicates that an agent which specifically decreases brain NA can attenuate self-stimulation. Roll (1970) showed that after disulfiram, rats had many pauses of greater than 60 s between lever-presses for brain-stimulation reward. The animals appeared to be sleepy in the long interpress intervals. She interpreted this as showing that disulfiram decreases self-stimulation by making animals drowsy. In the experiments of Wise and Stein (1969) it was also noted that the injections of disulfiram and the similar agent diethyldithiocarbamate (DEDTC) which attenuated self-stimulation produced sedation, and that reversal of the attenuation by intraventricular injections of L-norepinephrine also rapidly produced a state of arousal and alertness. Thus in these experiments it has not been shown that noradrenergic transmission mediates reward produced by brain stimulation. (The hypothesis under test is that the rewarding stimulation activates noradrenergic neurones, and that the release of NA which occurs for each bar-press mediates the reward.) It is possible that the effects on self-stimulation of the treatments are produced by a decrease in arousal, which is known to markedly affect self-stimulation rate (Rolls, 1971a, b, c; see section 3.7.4).

To investigate how noradrenergic and dopaminergic transmissions are involved in reward and arousal, Kelly *et al.* (1974) measured the effects on self-stimulation rate and two measures of sedation of disulfiram (which decreases the synthesis of NA but not DA), phentolamine (which blocks receptors sensitive to NA but not DA receptors—Nickerson and Hollenberg, 1967) and spiroperidol (which blocks DA but not NA receptors—Andén *et al.*, 1970). The level of sedation was measured by spontaneous locomotor activity, and by spontaneous rearing, a good measure of arousal/sedation (Benešová *et al.*, 1967; Cole and Dearnaley, 1971). Eight rats were tested while each of the drugs was active or after a placebo. A test consisted of a 5 min measurement of spontaneous locomotor activity and rearing followed by a 10 min test of lateral hypothalamic self-stimulation rate.

The bar histograms in Fig. 33 show that when disulfiram or phentolamine produced a modest reduction in self-stimulation rate (from 70 to about 42 bar-presses/min), the animals were very drowsy, as measured by the decrease in rearing and the decrease in locomotor activity. The animals also looked drowsy. Therefore inhibition of the synthesis of NA (disulfiram treatment) or blockade of noradrenergic receptors (phentolamine treatment) does reduce self-stimulation rate, but at the same time produces drowsiness. (The effect produced by the i.p. phentolamine is reproduced by the bilateral intracranial injection of 10 μg of phentolamine mesylate. The conclusion follows that the blockade of central noradrenergic receptors produces drowsiness, together with some reduction in self-stimulation rate.)

Whether the drowsiness accounts for the whole of the decrease in self-stimulation rate is not known. It is clear that the treatments by producing drowsiness would decrease self-stimulation rate (see section 3.7.4), and this type of experiment leaves open the question of

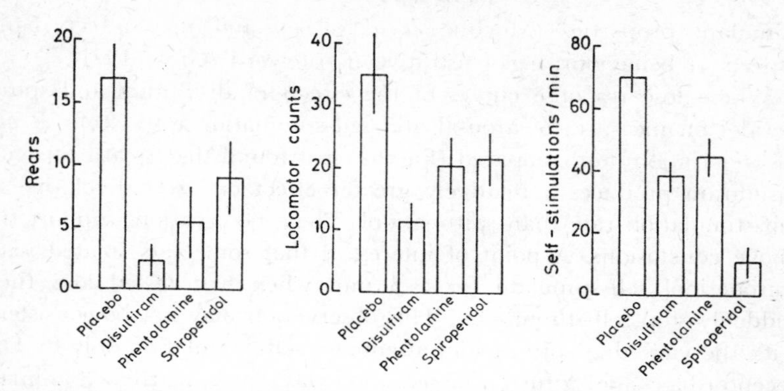

FIG. 33. Effect of disulfiram (200 mg/kg i.p. injected 3 h before testing), phentolamine mesylate (10 mg/kg i.p. injected 40 min before testing) and spiroperidol (0.1 mg/kg i.p. injected 2 h before testing) on two measures of arousal/sedation (rearing and locomotor activity) and on lateral hypothalamic self-stimulation rate. Eight rats were tested in counterbalanced order. Bars represent mean with SE of the mean indicated.

whether noradrenergic transmission plays any role in brain-stimulation reward other than by modifying arousal level. Although noradrenergic transmission could mediate reward, this present experiment suggests that such a role has never been proved. Rather, the effects obtained previously have probably been due at least partly to effects on arousal.

It is also shown in Fig. 33 that treatment with spiroperidol reduced self-stimulation rate from 70 to 10 stimulations/min, and left rearing and locomotor activity relatively high (compared with disulfiram and phentolamine). Therefore pharmacological treatments can be found which appear to attenuate reward aspects of self-stimulation more specifically with respect to arousal than disulfiram. It is interesting that spiroperidol produces a blockade of dopamine receptors (Andén *et al.*, 1970). This evidence therefore suggests that dopaminergic pathways are involved in brain-stimulation reward. Although sedation probably does normally reduce self-stimulation rate, Wise and Stein (1969) noted that some sedative drugs, e.g. barbiturates, may not decrease self-stimulation rate. Thus attenuation of self-stimulation by these drugs is probably more usual (Mogenson, 1964) and the facilitation sometimes seen may not be related to the sedation produced. These pharmacologically non-specific treatments have some stimulant properties (Machne *et al.*, 1955) and may affect many aspects of behaviour, e.g. frustrative non-reward (Gray, 1971).

When dose–response curves of the effects of disulfiram and spiroperidol on measures of arousal and self-stimulation rate (Rolls *et al.*, in preparation) are compared (Fig. 34) it is found that treatment with disulfiram produces a relatively greater effect on arousal relative to self-stimulation rate than spiroperidol. These observations support the above conclusions. A point of interest is that some rats treated with spiroperidol self-stimulate for 1–3 min when first tested and then suddenly stop self-stimulation. This observation may not be consistent with the view that spiroperidol attenuates self-stimulation only by DA receptor blockade. A further observation was that some treated animals then assumed a posture facing the lever and could not be induced to bar-press either by priming or by placing the animal on the lever. The hypothesis that spiroperidol attenuates self-stimulation by attenuating

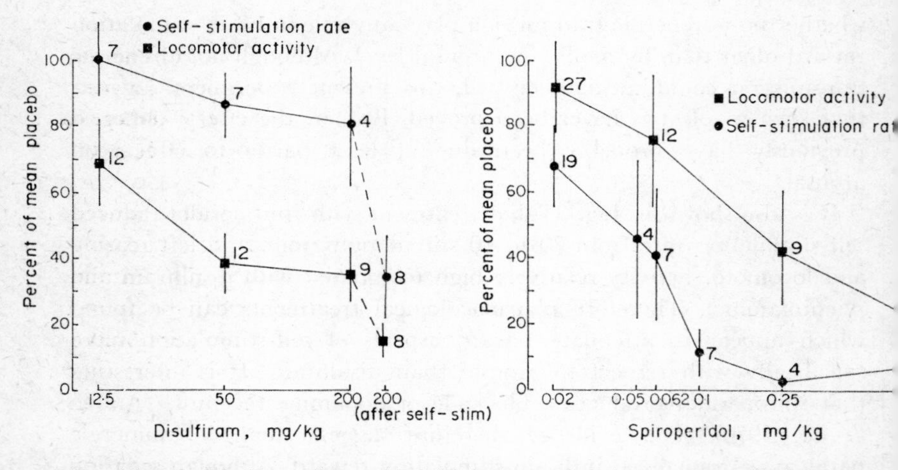

FIG. 34. *Left:* Disulfiram (injected i.p. 3 h before testing) has a greater effect on locomotor activity than on self-stimulation rate. One group of 8 rats was allowed to self-stimulate continuously following the disulfiram injection: after 3 h locomotor activity was also more depressed than the self-stimulation (see 200 mg/kg, after self-stimulation). *Right:* Spiroperidol (injected i.p. 2 h before testing) has a greater effect on self-stimulation rate than on locomotor activity. The numbers of animals tested are shown beside each point. The points represent the mean (± SE) locomotor activity or self-stimulation rate of the different rats, expressed as a percentage of the group's average locomotor activity or self-stimulation rate.

operant behaviour (although not general activity) rather than reward cannot be rejected.

The main conclusions of these studies therefore are as follows. Firstly, treatments which alter noradrenergic transmission and attenuate self-stimulation may do so at least partly by producing sedation. Secondly, pharmacological agents can be found which affect self-stimulation much more selectively. Thirdly, agents which produce DA-receptor blockade attenuate self-stimulation. Fourthly, DA-receptor blocking agents do not attenuate self-stimulation merely by producing sedation (compare dose–response curves for spiroperidol and disulfiram, Fig. 34), and may therefore affect brain-stimulation reward more specifically. Fifthly, although dopaminergic transmission could be involved in reward produced by brain stimulation, it has not been proved that spiroperidol (and pimozide) attenuate self-stimulation by blocking dopaminergic transmission in reward pathways.

There is now other evidence that DA is involved in self-stimulation of at least some sites. Crow *et al.* (1972) obtained self-stimulation when electrodes were near the DA-containing cell bodies (especially the group A10) in the ventral mesencephalon. Wauquier and Niemegeers (1972) showed that the DA-receptor blocking agent pimozide attenuates MFB self-stimulation. Rolls, Burton, and Shaw have recently extended the observations with spiroperidol to the monkey, showing that in the squirrel monkey a dose of 4 μg intracranially attenuates self-stimulation. Phillips and Fibiger (1973) found that D-amphetamine enhanced MFB self-stimulation 7–10 times more effectively than it enhanced self-stimulation of the substantia nigra. This provided an indication that noradrenergic synapses are involved in MFB self-stimulation and dopaminergic synapses in self-stimulation of the substantia nigra.

In a further investigation of the role of DA in self-stimulation it was found that spiroperidol attenuated self-stimulation of many different brain sites (the nucleus accumbens, septal area, anterior hypothalamus, and ventral tegmental area) (Kelly *et al.*, 1974; Rolls *et al.*, 1974b). Thus DA may be involved in self-stimulation of these different sites. The spiroperidol did not appear to produce its effect because of a simple motor impairment in these experiments, because at a particular dose of spiroperidol (e.g. 0.05 mg/kg) the rats could still bar-press quite fast (e.g. 40 presses/min for tegmental stimulation), yet showed an attenuation of nucleus accumbens self-stimulation below the 10 presses/min baseline. Thus DA-receptor blockade does not appear to attenuate self-stimulation because it limits how fast an animal can bar-press for the stimulation. With intracranial injections of spiroperidol it was found that self-stimulation was attenuated by smaller volumes in the preoptic area than in the caudate nucleus. This suggests that a structure near the preoptic area (e.g. the nucleus accumbens, a site of dopaminergic terminals) is more closely related to the blockade of self-stimulation than the caudate nucleus. Yet it was not possible to separate the effects of spiroperidol on motor behaviour (in particular catalepsy, measured by waxy flexibility—the tendency of an animal to maintain any position into which it is put—and time spent hanging motionless on a rod) from its effects on self-stimulation.

The conclusion which can be made at present then is that DA receptors are involved in self-stimulation but not necessarily in the reward produced by the stimulation. The role of DA receptors in self-stimulation could be in the high-level organization of the motor responses involved in self-stimulation, in line with the findings described above (see Rolls *et al.*, 1974a, b). However, none of the experiments described here rules out the possibility that DA (or even NA) is involved in reward produced by electrical stimulation of the brain.

In a different series of experiments (Rolls *et al.*, 1974b) it was found that DA-receptor blockade attenuates drinking less than eating, and eating less than self-stimulation, if the animals drank from a tube, ate pellets from the floor of their cages, or bar-pressed for brain-stimulation reward. But if the rats bar-pressed in a Skinner box to obtain water, or food, or brain-stimulation reward, then an equal and severe deficit of the drinking, eating, and self-stimulation was found. Thus one effect of DA-receptor blockade on motivated behaviour is to impair the motor responses made according to their complexity. It is not at all clear at present that there is any effect other than this motor function of the DA receptors in the motivation to drink, eat, or obtain brain-stimulation reward.

In a different type of experiment taken to support the noradrenergic theory of reward, Stein and Wise (1969) found that central NA turnover increased during MFB self-stimulation. The method involved perfusion of labelled NA precursor into the ventricles, and recovery of labelled NA from the brain. Although this is an interesting observation, it was not shown that the measured effect was due to reward rather than, for example, arousal, and there are difficulties with this type of experiment, e.g. substances may be released non-specifically (Chase and Kopin, 1968).

Another type of observation is cited in support of the noradrenergic theory of reward (Stein, 1969, 1971). It is held that self-stimulation sites are located on the course of ascending NA pathways, and that the stimulation of these pathways accounts at least in part for positive reinforcement. In line with this Crow (1972) and Ritter and Stein (1972) have found self-stimulation with electrodes in the region of the locus coeruleus. However, this evidence is inconclusive in that other different pathways also connect self-stimulation sites. For example, Routtenberg (1971) found that fibres coursed from reward sites in the medial prefrontal cortex in the rat through hypothalamic self-stimulation sites, and Rolls and Cooper (1973, 1974b) found that prefrontal neurones were activated in self-stimulation of many brain sites, including the region of the locus coeruleus. Further, Cooper and Rolls (1974) were not able to demonstrate powerful activation of neurones in the locus coeruleus during self-stimulation. Nevertheless, there is an impressive overlap between self-stimulation sites and the courses of the A9 and A10 dopaminergic and the A6 noradrenergic fibres (see review of German and Bowden, 1974). These authors suggest that excitation of any of these three, but not of the other, catecholamine fibre systems mediates self-stimulation. In line with this suggestion Clavier and Routtenberg (1974) did not find self-stimulation of the A1, A2, and A5 cell groups of origin of the ventral noradrenergic system. The present author does not find the overlap between catecholamine-containing fibres and self-stimulation conclusive evidence that catecholamine-containing fibres mediate brain-stimulation reward, and awaits further evidence on this fascinating problem.

Thus at present evidence supporting the noradrenergic theory of reward is not conclusive, and our acceptance or rejection of the theory must depend on future research. Dopamine is involved in self-stimulation, but whether it is involved in the reward produced by the stimulation (as opposed, for example, to the organization of motor responses) is unknown.

4.4. CATECHOLAMINES AND REWARD IN MAN

There have been extrapolations from the noradrenergic theory of reward to abnormal human emotional behaviour. In an extrapolation to depression, Stein (1967) starts from the position that the depressed patient despairs even when the environment supplies a normal amount of rewarding stimulation. The failure of rewarding stimulation to produce its usual effect led him to suggest that the reward mechanism of depressed patients is somehow deficient or pathologically hypoactive. Thus monoamine oxidase inhibitors would relieve depression by building up stores of NA in the presynaptic terminals of the reward synapses, while tricyclic antidepressants would compensate for deficient adrenergic transmission at the same synapses by retarding the re-uptake of NA after its release.

Another remarkably similar extrapolation is to schizophrenia (Stein and Wise, 1971; Stein, 1971). Following the observation that intraventricular injections of 6-hydroxydopamine in rats which lead to degeneration of catecholamine-containing pathways permanently attenuate self-stimulation, it was suggested that the lack of emotional responsiveness (lack of response to reward) found in schizophrenia could be produced as a result of degeneration of NA-containing reward pathways. The mechanism of degeneration postulated was the abnormal endogenous production of 6-hydroxydopamine. Chlorpromazine was held to have a therapeutic effect clinically because of its ability to prevent the uptake of 6-hydroxydopamine by the noradrenergic terminals. As the evidence for the noradrenergic theory of reward is itself weak, these speculative theories of emotional disorders must be treated with caution.

Nevertheless, if reward is important in emotional behaviour as suggested above, then research on the pharmacological basis of brain-stimulation reward is likely to be relevant to understanding the regulation of emotion. Certainly the view that catecholamines are involved in brain-stimulation reward is consistent with present views on the biochemistry of emotion (Schildkraut and Kety, 1967; Kety, 1970; Rech and Moore, 1971). Thus many treatments for disturbances of emotional behaviour are known to affect catecholamines, and alterations in catecholamine metabolism are seen in some emotional disorders. For example, one action of the major tranquillizers or neuroleptic drugs used to treat schizophrenia is to block NA and often more specifically DA receptors (Andén *et al.*, 1970). Antidepressant drugs such as monoamine oxidase inhibitors and imipramine are known to elevate brain levels of catecholamines (Schildkraut, 1965). L-Dopa, which is a precursor of DA and NA used to treat the akinesia (lack of voluntary movement) in Parkinson's disease (Hornykiewicz, 1973), may lead to excessive emotional behaviour (Sacks, 1973). Many of Sacks's patients reported that before treatment their behaviour was blocked. It may be that the effects of DA receptor blockade on reward (see above and Rolls *et al.*, 1974b) and punishment (see Wauquier and Niemegeers, 1972) may represent a similar interruption of voluntary behaviour. Thus many types of evidence do indicate a close link between emotion and catecholamine activity, and this deserves further investigation.

4.5. EFFECTS OF OTHER PHARMACOLOGICAL AGENTS ON SELF-STIMULATION

5-Hydroxytryptamine (serotonin) is contained in cells in the raphé nucleus of the midbrain which have axons which ascend in the ventral part of the medial forebrain bundle, and progress towards the septum, cingulum, amygdala, and cortex (see Moore, 1971; Ungerstedt, 1971). It may not be closely linked to brain-stimulation reward in that self-stimulation in the general region of the raphé nucleus and central gray of the midbrain is not affected by depletion of serotonin with *p*-chlorophenylalanine (Margules, 1969). Self-stimulation of the lateral hypothalamus in rats may be facilitated by *p*-chlorophenylalanine (500 mg/kg; Poschel and Ninteman, 1971), but pain sensitivity measured by the jump threshold is also increased (Tenen, 1967). Stark and Fuller (1972) found that *p*-chlorophenylalanine inhibited self-stimulation in rats and dogs. No clear conclusion seems possible.

Ascending cholinergic pathways have been described by Shute and Lewis (1966). The most dramatic effect of peripherally administered cholinergic antagonists such as scopolamine is to increase resistance to extinction. That is, self-stimulation (and food-rewarded behaviour) appears normally, but when the reward is no longer given, the animals continue to bar-press for very long periods (Olds, 1970; observations of Rolls and Shaw). Newman (1972) has found that the anticholinesterase physostigmine (0.2 mg/kg) attenuates lateral hypothalamic self-stimulation in the rat. The effect is central in that it is blocked by atropine but not by atropine methylnitrate which does not cross the blood–brain barrier. Thus a muscarinic (cholinergic) system can attenuate self-stimulation. He also postulated a nicotinic (cholinergic) system which facilitates self-stimulation. The observations that blockade of a cholinergic (muscarinic) system disinhibits responding in extinction and facilitates self-stimulation may be compared with the disinhibition of punished behaviour produced by cholinergic blockade in the ventromedial hypothalamus (Stein, 1969). Treated rats showed a passive avoidance deficit, measured as continued bar-pressing for food even when this was accompanied by shock. Carlton (1969) has argued that cholinergic systems may be involved in many types of behavioural inhibition.

In addition to their attenuation of rewarded behaviour (see above), the major tranquillizers or neuroleptic drugs also attenuate punished behaviour (see Margules and Stein, 1967; Wauquier and Niemegeers, 1972). In contrast, minor tranquillizers or anti-anxiety (anxiolytic) drugs attenuate the effects of punishment but not of reward on behaviour (Margules and Stein, 1967). For example, at a site where electrical stimulation produced both reward and aversion, oxazepam (a minor tranquillizer) increased the number of times the rat pressed for the stimulation.

4.6. SUMMARY

There is considerable evidence that central catecholamines are involved in intracranial self-stimulation. Much of the evidence which has been cited to show that the release of the particular catecholamine, noradrenaline, mediates brain-stimulation reward is weak in two respects. Firstly, many of the treatments which decrease self-stimulation rate (e.g. α-methyl-*p*-tyrosine, chlorpromazine, haloperidol) or increase self-stimulation rate (e.g. amphetamine) affect DA as well as NA. Secondly, most of the treatments affect arousal, and it has not been shown, for example, that sedation does not account for the attenuation of self-stimulation produced by agents which decrease the synthesis of NA. Thus the evidence available at present does not provide adequate support for the view that the release of NA mediates brain-stimulation reward. Dopamine-receptor blockade attenuates self-stimulation of many different brain sites without producing major sedation. Whether this represents a true block of reward or is a result of interference with high-level motor behaviour is not yet clear. Suggestions that, in man, hypoactivity in a noradrenergic reward system is involved in depression, and that degeneration of a noradrenergic reward system is involved in the hypoemotionality of schizophrenia, have been made and require critical investigation. It is probable that neither serotonin nor acetylcholine plays a primary role in brain-stimulation reward.

CHAPTER 5

POSSIBLE ANOMALOUS EFFECTS OF BRAIN-STIMULATION REWARD

BRAIN-STIMULATION reward has sometimes been felt to have effects which are unlike those of normal rewards. In many cases there are now explanations of these apparent anomalies. The investigations into the properties of brain-stimulation reward have often clarified the properties of natural rewards.

5.1. PRIMING

Rats will run along a runway to obtain electrical stimulation at the end of the runway. If some of the electrical stimulation is given in addition at the start of the runway, then the rats will run faster (Gallistel, 1969a). The stimulation given at the start of the runway is said to have a priming effect, which decays gradually (Fig. 35—the priming effect decayed over a particularly long time in this rat). Priming stimulation (given by the experimenter) may also cause a rat to return to a self-stimulation lever and start self-stimulating. This may be useful if the rat does not bar-press at once when it is put in a self-stimulation box, i.e. if it shows an "overnight decrement effect".

The priming effect is similar in some ways to a natural drive. This was shown by Deutsch *et al.* (1964), who gave thirsty rats a choice of water or brain-stimulation reward. If the animals were given priming stimulation immediately before making the choice, they chose brain-stimulation reward rather than water (Fig. 36, 0 s delay; the 5 h water deprivation condition is the simplest). This priming effect decayed over about 60 s, as shown by the decreasing preference for

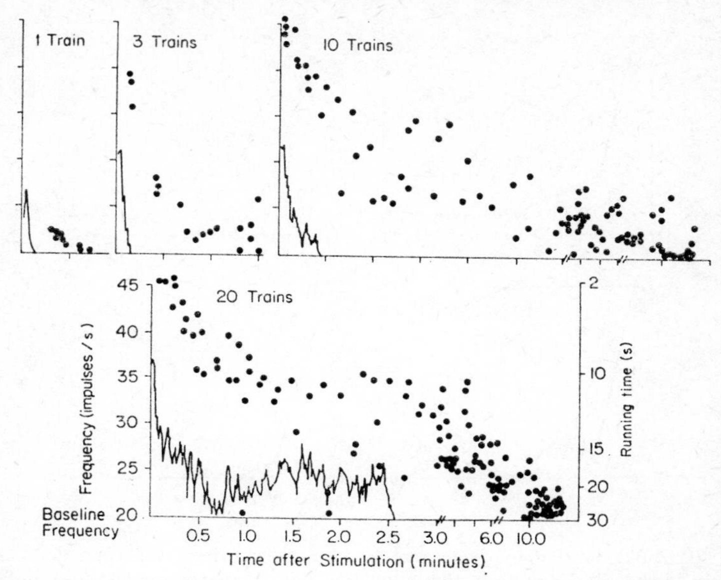

FIG. 35. Running speed (closed circles) for a small brain-stimulation reward on massed trials as a function of time after different numbers of priming trains of stimulation given before the first trial. The abscissa has been shortened (data from Gallistel, *J. Comp. Physiol. Psychol.* **69,** 713–21, 1969). The firing rate of an arousal unit (solid line) as a function of time after different numbers of trains of stimulation delivered to a MFB self-stimulation site is shown for comparison.

brain-stimulation reward as opposed to water. The priming effect is like a natural drive in that it competes with a natural drive such as thirst.

Gallistel (1969b) tested whether the priming effect was like a specific natural drive, equivalent, for example, to hunger at one stimulation site and thirst at another site. He did not find that priming stimulation at one site affected the subsequent choice of brain-stimulation reward site differently from priming at other sites. Thus either many specific natural drives were elicited at all the sites, or the priming effect is a rather general type of drive-like effect. The latter interpretation is the more parsimonious.

A model of the priming effect has been put forward by Deutsch (1960) and discussed by Gallistel (1964, 1973). A train of stimulation

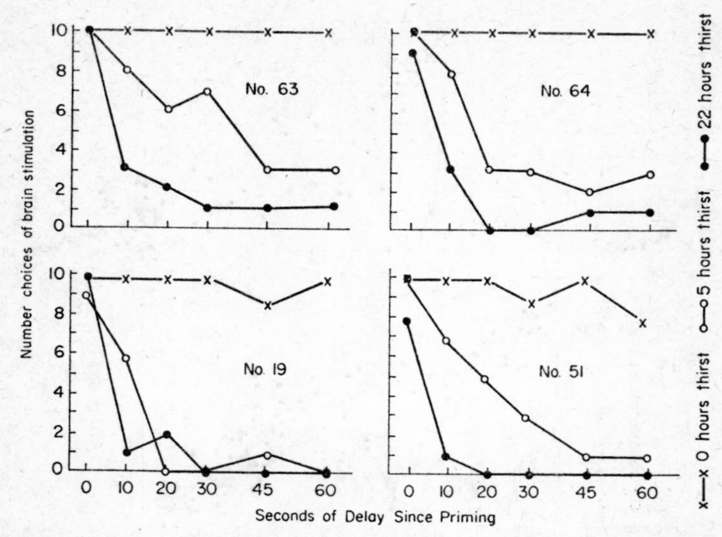

FIG. 36. Choice behaviour of four rats as a function of degree of water deprivation and time since priming at a self-stimulation site. The number of times brain-stimulation reward is chosen rather than water decreases for about 60 s following the priming (see, for example, curves for 5 h, water deprivation). This appears to show the decay of a specific drive-like process. (From Deutsch *et al.*, *J. Comp. Physiol. Psychol.* **57,** 241–3, 1964.)

at a self-stimulation site is supposed to have two effects. The first is a reward effect, which provides the reward for bar-pressing. The second is a separate effect like a specific natural drive (e.g. hunger or thirst) which provides the drive or motivation for more brain-stimulation reward. This drive is supposed to decay gradually over a period of seconds or minutes, and thus accounts for the "time elapsed since last stimulation" phenomena, e.g. the need for priming after withdrawal of reward, the overnight decrement effect, the inter-trial interval effect, and the poor performance on low density reward schedules. This theory may not be necessary, as there are a number of partial or complete explanations of the priming phenomena, as shown below.

Firstly, it is known that arousal, measured by EEG desynchronization and by increased locomotor activity, is produced by stimulation of reward sites along the medial forebrain bundle (MFB) (Rolls,

1971a, c; Rolls and Kelly, 1972; see Chapter 3). This arousal decays in a very similar way to the priming effect, as shown, for example, in Fig. 35. The arousal must at least affect MFB self-stimulation. Because of its similarities to the priming effect, it may well mediate the priming effect, at least in part. Similarities include the temporal nature of the decay, the greater magnitudes and durations of the effects produced by more stimulation trains, and the refractory periods of the neurones through which the effects are produced (Rolls, 1971a, b, 1974). A further similarity is that although the priming effect can be distinguished from and competes with a natural drive such as thirst (Deutsch *et al.*, 1964, see above), the priming effects produced at different MFB sites appear to be similar to each other (Gallistel, 1969b, see above). In addition to providing at least a partial explanation for the priming effect, arousal, which is elicited by rewarding stimulation of MFB but not nucleus accumbens sites, may account for some of the differences between MFB and rhinencephalic self-stimulation (Rolls, 1971c). Such differences include hyperactivity and fast rates of self-stimulation seen with MFB stimulation (see Rolls, 1971c, 1974).

Secondly, incentive motivation (see section 2.1.3) may account for some priming phenomena. For example, if satiated animals are primed with an intraoral reward of chocolate milk, then they will resume bar-pressing to obtain more chocolate milk (Panksepp and Trowill, 1967b). It is important that this effect is seen most markedly under zero-drive conditions, e.g. when an animal is neither hungry nor thirsty. These are the conditions under which self-stimulation experiments are often run. Incentive motivation is seen in other situations with natural reward, e.g. as the salted-nut phenomenon (Hebb, 1949), and as the increase in the rate of ingestion seen at the start of a meal (Le Magnen, 1971).

Thirdly, conflict may account for some priming phenomena (Kent and Grossman, 1969). These experiments showed that only some rats needed priming after an interval in which self-stimulation was prevented by withdrawal of the lever. In the "primer" rats the stimulation seemed to produce reward and aversion in that self-stimulation was accompanied by squeaking, defecation, and cringing. It was suggested that during the time away from the bar the reward decayed more rapidly than the aversion, so that self-stimulation was not resumed without priming. It was also found that "non-primer" rats could be converted into "primers" by pairing painful tail shock with the brain-stimulation reward. Although Kent and Grossman labelled one group of their rats as "non-primers", a priming effect can be demonstrated in this type of animal using, for example, a runway situation (Reid *et al.*, 1973).

Thus there are several explanations of the priming effect which are not necessarily contradictory and which may all contribute to the priming effect.

5.2. EXTINCTION

Extinction from self-stimulation can be very rapid (Seward *et al.*, 1959) even when the manipulandum is withdrawn, so that non-rewarded responses are prevented (Howarth and Deutsch, 1962). A factor which determines the number of responses in extinction is the time elapsed since the last reward (Howarth and Deutsch, 1962) which was not found to be true for thirsty rats trained for water (Quartermain and Webster, 1968). Perhaps the major factor which accounts for the rapid extinction of self-stimulation often reported is that no relevant drive is present. If rats are hungry, extinction from brain-stimulation reward is very prolonged (Deutsch and Di Cara, 1967). Thus one conclusion which can be made from studies with brain-stimulation reward is that resistance to extinction depends on the presence of an appropriate drive. This has probably been demonstrated for conventional reinforcers by Panksepp and Trowill (1967b), who found a trend towards rapid extinction in satiated rats previously rewarded with chocolate milk (insufficient rats were run for a definite conclusion—Gallistel, 1973). Another factor which contributes to rapid extinction is the short time between pressing the manipulandum and receiving brain-stimulation reward, in contrast to the one or two seconds delay between pressing a manipulandum and then finding food or water reward available in a cup (Gibson *et al.*, 1965; see also Panksepp and Trowill, 1967a). Keesey (1964) has also found that errors in learning a brightness discrimination are increased when the brain-stimulation reward is delayed.

5.3. SECONDARY REINFORCEMENT

Another feature of brain-stimulation reward which was thought to be anomalous is secondary reinforcement. Seward *et al.* (1959) and Mogenson (1965) had difficulty in obtaining secondary reinforcement with brain-stimulation reward. When a previously neutral stimulus was paired with brain-stimulation reward, the stimulus did not come to acquire reinforcing properties. Di Cara and Deutsch (cited in Deutsch and Deutsch, 1966, p. 128) found that secondary reinforcement occurred normally if the rats had a relevant drive, e.g. hunger or thirst. Thus brain-stimulation reward has helped to clarify conditions under which secondary reinforcement occurs.

5.4. PREFERENCE FOR INTERMITTENT REWARD

If rewarding stimulation is connected so that it is on while a lever is pressed, animals will repeatedly press and release the lever. Cats will even learn one response to switch the stimulation on and another to switch it off, and alternate these repeatedly (Roberts, 1958). The latency of both the escape and the approach responses decreases as the intensity of the stimulation increases (Valenstein and Valenstein, 1964; Hodos, 1965). Do these observations mean that brain-stimulation reward produces aversion if it is prolonged? This is probably not the case. Hodos (1965) reported that rats that work to turn stimulation on and off repeatedly will work harder to obtain long as opposed to short trains of stimulation. Kelly (personal communication, 1970) found that rats allowed only one train of brain-stimulation reward per day gradually learned to leave it on longer. Deutsch and Hawkins (1972) found that after rats had been receiving brain-stimulation reward for a short period the rats chose to increase its intensity and did not choose to terminate it. Thus the stimulation had not become aversive, but rather the reward value of it seemed to have decreased. This adaptation to reward suggests that rats terminate brain-stimulation reward so that they can switch it on again. Nevertheless, it seems very likely that at sites between reward and aversion areas mixed reward and aversion will be produced by electrical stimulation.

5.5. ANALGESIA

In man, stimulation of some rhinencephalic (Brady, 1961) and medial forebrain bundle (Ervin *et al.*, 1969) reward sites can provide relief from pain or anxiety. Pain relief produced by MFB stimulation rises to a maximum 15 min after the stimulation, and may last several hours. Stimulation of the amygdala can give release of tension for several hours (Ervin *et al.*, 1969). In the rat rewarding rhinencephalic (septal) stimulation suppresses the aversive effects of tegmental stimulation (Routtenberg and Olds, 1963), and rewarding hypothalamic stimulation can attenuate the aversive properties of peripheral shock (Cox and Valenstein, 1965). Keene and Casey (1973) have recently found that medial thalamic units show opposite responses to rewarding and aversive brain stimulation.

Apparent inconsistencies are that rewarding MFB stimulation augments the behavioural response to aversive stimulation (Olds and Olds, 1962; Stein, 1965), and rewarding septal stimulation reduces escape from midbrain stimulation, but facilitates escape from foot shock (Gardner and Malmo, 1969).

5.6. REINFORCEMENT

Only recently has intracranial self-stimulation been used as a tool to investigate reinforcement, i.e. the learning which occurs when a reinforcer is applied. A reinforcer is an event, stimulus, or state of affairs that changes subsequent behaviour when it temporally follows an instance of that behaviour. In one study the locus coeruleus was lesioned bilaterally to investigate its role in learning in view of its suspected role in brain-stimulation reward (Anzelark *et al.*, 1973). The lesioned rats showed slow learning to run along an L-shaped runway to obtain food reward. Clearly many factors could have diminished the rate of increase of running in the lesioned rats. Therefore a number of control experiments were run, and it was reported that the lesioned rats had a normal weight gain, normal motor and exploratory activity in an open field, and could discriminate sucrose from water. Nevertheless, other investigators report that lesions in this region affect the sleep–waking cycle. For example, lesions in the locus coeruleus in the cat may lead (depending on the exact lesion site) to a decrease in paradoxical sleep (measured by muscle tone and/or lack of EEG desynchronization), inattention to auditory and visual stimuli, and hallucinatory behaviour (Jouvet, 1972). It follows that further investigation will be necessary to show whether there is a specific deficit in learning after locus coeruleus lesions. Comparably careful investigations will also be needed to determine whether lesions of the A9 and A10 DA-containing neurones affect learning, because lesions in this area produce general changes in behaviour such as drowsiness (Jouvet, 1972; Jones *et al.*, 1973).

The studies which implicate the amygdala in stimulus–reinforcement learning (see section 3.7.2) provide a good example of how it has recently been possible to analyse the neural basis of learning. Lesions of the amygdala attenuate the formation of stimulus–reinforcement associations, electrical stimulation of the amygdala can provide reward and punishment, and single units in the amygdala appear to show stimulus-reinforcement association learning. Neuroanatomically the amygdala is well placed for participation in this process in that it receives highly analysed sensory inputs, and has connections with hypothalamic areas which are implicated in basic reward and punishment processes. Whether the stimulus–reinforcement learning of hypothalamic neurones (see Chapters 2 and 3 and Olds, 1973) is dependent on the amygdala remains to be shown.

Thus investigation of the relation between brain-stimulation reward and reinforcement is starting to clarify the neural basis of learning and of emotion.

5.7. SUMMARY

There are now explanations of a number of the apparently anomalous aspects of self-stimulation. The explanations have often clarified the nature of natural rewards. The priming effect, in which rewarding brain stimulation may facilitate subsequent performance to obtain more brain stimulation, is seen with natural rewards, particularly under the zero-drive conditions (no hunger, thirst, etc.) typical of self-stimulation experiments. Incentive motivation, arousal produced by the brain stimulation, and conflict when the stimulation is both rewarding and aversive may contribute to the priming effect. Factors important in determining the rate of extinction from brain-stimulation (and natural) reward are presence or absence of a relevant drive such as hunger, and the delay between making a response and obtaining a reward. Secondary reinforcement occurs with brain-stimulation reward if a relevant drive (such as hunger) is present. Adaptation of reward neurones to continuous stimulation or aversion produced by the stimulation may account for the preference for intermittent brain-stimulation reward. It is not known how the analgesia produced by stimulation at some reward sites is related to brain-stimulation reward. Because of its close relation to learning, brain-stimulation reward is now being used to analyse the neural basis of learning. Neurones in the amygdala appear to provide a basis for learning to associate a stimulus with reward (or punishment), and by allowing learning about what is attractive or aversive amygdaloid neurones are crucial in motivational and emotional behaviour. It has also been suggested that noradrenergic and dopaminergic neurones are involved in learning.

POSTSCRIPT

THE summaries which follow each chapter indicate some of the conclusions which can be made about brain-stimulation reward and from studies using brain-stimulation reward. I hope it is clear that our understanding of brain-stimulation reward is progressing rapidly at present, and is clarifying such diverse problems as the controls of eating, drinking, and sexual behaviour, the basic processes involved in emotional behaviour, the function of complex brain regions such as the hypothalamus, limbic structures, and the prefrontal cortex, the pharmacology of normal and abnormal behaviour, and also the neural basis of learning.

REFERENCES

ANDÉN, N. E., BUTCHER, S. G., CORRODI, H., FUXE, K., and UNGERSTEDT, U. (1970) Receptor activity and turnover of dopamine and noradrenaline after neuroleptics. *Eur. J. Pharmac.* **11,** 303–14.

ANZELARK, G. M., CROW, T. J., and GREENAWAY, A. P. (1973) Impaired learning and decreased cortical norepinephrine after bilateral locus coeruleus lesions. *Science* **181,** 682–4.

ASDOURIAN, D., STUTZ, R. M., and ROCKLIN, K. W. (1966) Effects of thalamic and limbic system lesions on self-stimulation. *J. Comp. Physiol. Psychol.* **61,** 468–72.

BAGSHAW, M. H. and COPPOCK, H. W. (1968) Galvanic skin response conditioning deficit in amygdalectomized monkeys. *Expl. Neurol.* **20,** 188–96.

BALAGURA, S. and HOEBEL, B. G. (1967) Self-stimulation of the hypothalamic "feeding-reward system" modified by insulin and glucagon. *Physiol. Behav.* **2,** 337–40.

BALL, G. G. (1967) Electrical self-stimulation of the brain and sensory inhibition. *Psychon. Sci.* **8,** 489–90.

BALL, G. G. (1974) Vagotomy: effect on electrically elicited eating and self-stimulation in the lateral hypothalamus. *Science* **184,** 484–5.

BALL, G. G. and GRAY, J. A. (1971) Septal self-stimulation and hippocampal activity. *Physiol. Behav.* **6,** 547–9.

BARNETT, S. A. (1958) Experiments on "neophobia" in wild and laboratory rats. *Br. J. Psychol.* **49,** 195–201.

BEN ARI, Y. and LE GAL LA SALLE, G. (1972) Plasticity at unitary level: II, Modifications during sensory–sensory association procedures. *Electroenceph. Clin. Neurophysiol.* **32,** 667–79.

BENEŠOVÁ, O., ŠIMÁNĚ, Z., and KUNZ, K. (1967) Pyruvate, alpha-keto-glutarate and gamma-aminobutyrate in brains of rats with different levels of excitability. *Physiol. Behav.* **2,** 203–5.

BERGQUIST, E. H. (1970) Output pathways of hypothalamic mechanisms for sexual, aggressive and other motivated behaviors in opposum. *J. Comp. Physiol. Psychol.* **70,** 389–98.

BISHOP, M. P., ELDER, S. T., and HEATH, R. G. (1963) Intracranial self-stimulation in man. *Science* **140,** 394–5.

BLASS, E. M. and EPSTEIN, A. N. (1971) A lateral preoptic osmosensitive zone for thirst in the rat. *J. Comp. Physiol. Psychol.* **76,** 378–94.

BOGACZ, J., ST. LAURENT, J., and OLDS, J. (1965) Dissociation of self-stimulation and epileptiform activity. *Electroenceph. Clin. Neurophysiol.* **19,** 75–87.

BOYD, E. S. and GARDNER, L. C. (1962) Positive and negative reinforcement from intracranial self-stimulation in teleosts. *Science* **136,** 648.

BOYD, E. S. and GARDNER, L. C. (1967) Effect of some brain lesions on intracranial self-stimulation in the rat. *Am. J. Physiol.* **213,** 1044–52.

Brady, J. V. (1960) Temporal and emotional effects related to intracranial electrical self-stimulation. In *Electrical Studies on the Unanesthetized Brain* (ed. E. R. Ramey and D. S. O'Doherty), Hoeber, New York.
Brady, J. V. (1961) Motivational–emotional factors and intracranial self-stimulation. In *Electrical Stimulation of the Brain* (ed. D. E. Sheer), University of Texas Press, Austin.
Brady, J. V. and Conrad, D. G. (1960) Some effects of limbic system self-stimulation upon conditioned emotional behavior. *J. Comp. Physiol. Psychol.* **53,** 128–37.
Brodal, A. (1969) *Neurological Anatomy,* 2nd edn., Oxford University Press, London.
Brodie, D. A., Moreno, O. M., Malis, J. L., and Boren, J. J. (1960) Rewarding properties of intracranial stimulation. *Science* **131,** 929–30.
Bruner, A. (1966) Facilitation of classical conditioning in rabbits by reinforcing brain stimulation. *Psychon. Sci.* **6,** 211–12.
Burton, M. J. and Rolls, E. T. Neurophysiological convergence of natural and brain-stimulation reward. Paper in preparation.
Butter, C. M. (1969) Perseveration in extinction and in discrimination reversal tasks following selective frontal ablations in *Macaca mulatta. Physiol. Behav.* **4,** 163–71.
Cabanac, M. (1971) Physiological role of pleasure. *Science* **173,** 1103–7.
Cabanac, M., Duclaux, R., and Spector, N. H. (1971) Sensory feedback in regulation of body weight: is there a ponderostat? *Nature* **229,** 125–7.
Caggiula, A. R. (1967) Specificity of copulation-reward systems in the posterior hypothalamus. *Proceedings of the 75th Convention, American Psychological Association,* pp. 125–6.
Caggiula, A. R. (1970) Analysis of the copulation-reward properties of posterior hypothalamic stimulation in male rats. *J. Comp. Physiol. Psychol.* **70,** 399–412.
Caggiula, A. R. and Hoebel, B. G. (1966) A "copulation-reward site" in the posterior hypothalamus. *Science* **153,** 1284–5.
Cantor, M. B. and LoLordo, V. M. (1970) Rats prefer signaled reinforcing brain stimulation to unsignaled ESB. *J. Comp. Physiol. Psychol.* **71,** 183–91.
Carlton, P. L. (1969) Brain-acetylcholine and inhibition. In *Reinforcement and Behavior,* ch. 10, pp. 266–327 (ed. J. T. Tapp), Academic Press, New York.
Chase, T. N. and Kopin, I. J. (1968) Stimulus-induced release of substances from olfactory bulb using the push–pull cannula. *Nature* **217,** 466–7.
Clavier, R. M. and Routtenberg, A. (1974) Ascending monoamine-containing fiber pathways related to intracranial self-stimulation: histochemical fluorescence study. *Brain Res.* **72,** 25–40.
Cole, J. and Dearnaley, D. P. (1971) A technique for measuring exploratory activity in rats: some effects of chlorpromazine and chlordiazepoxide. *Arzneimittel-Forsch.* (*Drug Res.*) **21,** 1359–62.
Coons, E. E. and Cruce, J. A. F. (1968) Lateral hypothalamus: food current intensity in maintaining self-stimulation. *Science* **159,** 1117–19.
Coons, E. E. and Fonberg, E. (1963) Paper read at EPA meeting in New York, cited in Glickman, S. E. and Schiff, B. B. (1967) A biological theory of reinforcement. *Psychol. Rev.* **74,** 81–109.
Cooper, R. M. and Taylor, L. H. (1967) Thalamic reticular system and central grey: self-stimulation. *Science* **156,** 102–3.
Cooper, S. J. and Rolls, E. T. (1974) Relation of activation of neurones in the pons and medulla to brain-stimulation reward. *Expl. Brain Res.* **20,** 207–22.

Cowey, A. and Gross, C. G. (1970) Effects of foveal prestriate and inferotemporal lesions on visual discrimination by rhesus monkeys *Expl. Brain Res.* **11,** 128–44.

Cox, V. C. and Valenstein, E. S. (1965) Attenuation of aversive properties of peripheral shock by hypothalamic stimulation. *Science* **149,** 323–5.

Crow, T. J. (1972) A map of the rat mesencephalon for electrical self-stimulation. *Brain Res.* **36,** 265–73.

Crow, T. J., Spear, P. J., and Arbuthnott, G. W. (1972) Intracranial self-stimulation with electrodes in the region of the locus coeruleus. *Brain Res.* **36,** 275–87.

Dahlström, A. and Fuxe, K. (1965) Evidence for the existence of monoamine-containing neurons in the central nervous system: I, Demonstration of monoamines in the cell bodies of brain stem neurons. *Acta physiol. scand.* **62,** suppl. 232, 1–55.

Delgado, J. M. R. (1960) Emotional behavior in animals and humans. *Psychiat. Res. Rep.* **12,** 259–71.

Delgado, J. M. R. (1969) Physical control of the mind, *World Perspectives*, vol. xli (ed. R. N. Anshen), reprinted 1971 by Harper & Row, New York.

Delgado, J. M. R. and Hamlin, H. (1960) Spontaneous and evoked electrical seizures in animals and in humans. In *Electrical Studies on the Unanesthetized Brain*, pp. 133–58 (ed. E. R. Ramey and D. S. O'Doherty), Hoeber, New York.

Deutsch, J. A. (1960) *The Structural Basis of Behavior*, University of Chicago Press, Chicago.

Deutsch, J. A. and Deutsch, D. (1966) *Physiological Psychology*, Dorsey Press, Homewood, Ill.

Deutsch, J. A. and Deutsch, D. (1973) *Physiological Psychology*, 2nd edn., Dorsey Press, Homewood, Ill.

Deutsch, J. A. and Di Cara, L. (1967) Hunger and extinction in intracranial self-stimulation. *J. Comp. Physiol. Psychol.* **63,** 344–7.

Deutsch, J. A. and Hawkins, R. D. (1972) Adaptation as a cause of apparent aversiveness of prolonged rewarding brain stimulation. *Behav. Biol.* **7,** 285–90.

Deutsch, J. A., Adams, D. W., and Metzner, R. J. (1964) Choice of intracranial stimulation as a function of delay between stimulations and strength of competing drive. *J. Comp. Physiol. Psychol.* **57,** 241–3.

Eleftheriou, B. E. (ed.) (1972) *The Neurobiology of the Amygdala*, Plenum Press, New York.

Epstein, A. N. (1960) Reciprocal changes in feeding behavior produced by intrahypothalamic chemical injections. *Am. J. Physiol.* **199,** 969–74.

Epstein, A. N. (1967) Oropharyngeal factors in feeding and drinking. In *Handbook of Physiology*, Section 6, *Alimentary Canal:* Vol. I, *Food and Water Intake*, pp. 197–218, American Physiological Society, Washington.

Ervin, F. R., Mark, V. H., and Stevens, J. (1969) Behavioral and affective responses to brain stimulation in man. In *Neurobiological Aspects of Psychopathology*, pp. 54–65 (ed. J. Zubin and C. Shagass), Grune & Stratton, New York.

Freeman, W. J. and Watts, J. W. (1950) *Psychosurgery in the Treatment of Mental Disorders and Intractable Pain*, 2nd edn., Thomas, Springfield, Ill.

Fulton, J. F. (1951) *Frontal Lobotomy and Affective Behavior. A Neurophysiological Analysis*, W. W. Norton, New York.

Fuster, J. M. and Uyeda, A. A. (1971) Reactivity of limbic neurons of the monkey to appetitive and aversive signals. *Electroenceph. Clin. Neurophysiol.* **30,** 281–93.

Gallistel, C. R. (1964) Electrical self-stimulation and its theoretical implications. *Psychol. Bull.* **61,** 23–34.

GALLISTEL, C. R. (1969a) The incentive of brain-stimulation reward. *J. Comp. Physiol. Psychol.* **69,** 713–21.

GALLISTEL, C. R. (1969b) Self-stimulation: failure of pretrial stimulation to affect rats' electrode preference. *J. Comp. Physiol. Psychol.* **69,** 722–9.

GALLISTEL, C. R. (1973) Self-stimulation: the neurophysiology of reward and motivation. In *The Physiological Basis of Memory,* ch. 7, pp. 175–267 (ed. J. A. Deutsch), Academic Press, New York.

GALLISTEL, C. R. and BEAGLEY, G. (1971) Specificity of brain-stimulation reward in the rat. *J. Comp. Physiol. Psychol.* **76,** 199–205.

GARDNER, L. and MALMO, R. B. (1969) Effects of low-level septal stimulation on escape: significance for limbic-midbrain interactions in pain. *J. Comp. Physiol. Psychol.* **68,** 65–73.

GERMAN, D. C. and BOWDEN, D. M. (1974) Catecholamine systems as the neural substrate for intracranial self-stimulation: a hypothesis. *Brain Res.* **73,** 381–419.

GIBSON, W. E., REID, L. D., SAKAI, M., and PORTER, P. B. (1965) Intracranial reinforcement compared with sugar–water reinforcement. *Science* **148,** 1357–9.

GOLDMAN, P. S., ROSVOLD, H. E., VEST, B., and GALKIN, T. W. (1971) Analysis of the delayed-alternation deficit produced by dorso-lateral pre-frontal lesions in the rhesus monkey. *J. Comp. Physiol. Psychol.* **77,** 212–20.

GOLDSTEIN, M. and NAKAJIMA, K. (1967) The effect of disulfiram on catecholamine levels in the brain. *J. Pharmac. Exp. Ther.* **157,** 96–102.

GOLDSTEIN, R., HILL, S. Y., and TEMPLER, D. I. (1970) Effect of food deprivation on hypothalamic self-stimulation in stimulus-bound eaters and non-eaters. *Physiol. Behav.* **5,** 915–18.

GOODMAN, I. J. and BROWN, J. L. (1966) Stimulation of positively and negatively reinforcing sites in the avian brain. *Life Sci.* **5,** 693–704.

GRASTAYÁN, E. and ÁNGYÁN, L. (1967) The organisation of motivation at the thalamic level of the cat. *Physiol. Behav.* **2,** 5–13.

GRASTAYÁN, E., KARMOS, G., VERECZKEY, L., MARTIN, J., and KELLENYI, L. (1965) Hypothalamic motivational processes as reflected by their hippocampal electrical correlates. *Science* **149,** 91–93.

GRAY, J. A. (1971) *The Psychology of Fear and Stress,* Weidenfeld & Nicholson, London.

GROSS, C. G. (1973) Visual functions of inferotemporal cortex. In *Handbook of Sensory Physiology,* vol. 7, part 3 (ed. R. Jung), Springer, Berlin.

GROSSMAN, S. P. (1973) *Essentials of Physiological Psychology,* Wiley, New York.

HEATH, R. G. (1954) *Studies in Schizophrenia. A Multidisciplinary Approach to Mind–Brain Relationships,* Harvard University Press, Cambridge.

HEATH, R. G. (1963) Electrical self-stimulation of the brain in man. *Am. J. Psychiat.* **120,** 571–7.

HEATH, R. G. (1964) Pleasure response of human subjects to direct stimulation of the brain: physiologic and psychodynamic considerations. In *The Role of Pleasure in Behavior,* pp. 219–43 (ed. R. G. Heath), Hoeber, New York.

HEATH, R. G. (1972) Pleasure and brain activity: deep and surface encephalograms during orgasm. *J. Nerv. Ment. Dis.* **154,** 3–18.

HEATH, R. G. and MICKLE, W. A. (1960) Evaluation of seven years' experience with depth electrode studies in human patients. In *Electrical Studies on the Unanesthetized Brain* (ed. E. R. Ramey and D. S. O'Doherty), Hoeber, New York.

HEBB, D. O. (1949) *The Organization of Behavior: A Neuropsychological Theory,* Wiley, New York.

HERBERG, L. J. (1963) Seminal ejaculation following positively reinforcing electrical stimulation of the rat hypothalamus. *J. Comp. Physiol. Psychol.* **56,** 679–85.

HIGGINS, J. W., MAHL, G. F., DELGADO, J. M. R., and HAMLIN, H. (1956) Behavioral changes during intracerebral electrical stimulation. *Archs. Neurol. Psychiat. Chicago* **76,** 399–419.

HILL, A. V. (1936) Excitation and accommodation in nerve. *Proc. R. Soc.* B, **119,** 305–55.

HODOS, W. (1965) Motivational properties of long durations of rewarding brain stimulation. *J. Comp. Physiol. Psychol.* **59,** 219–24.

HODOS, W. and VALENSTEIN, E. S. (1962) An evaluation of response rate as a measure of rewarding intracranial stimulation. *J. Comp. Physiol. Psychol.* **55,** 80–84.

HOEBEL, B. G. (1967) Intragastric balloon without gastric surgery for the rat. *J. Appl. Physiol.* **22,** 189–90.

HOEBEL, B. G. (1968) Inhibition and disinhibition of self-stimulation and feeding: hypothalamic control and postingestional factors. *J. Comp. Physiol. Psychol.* **66,** 89–100.

HOEBEL, B. G. (1969) Feeding and self-stimulation. *Ann. NY Acad. Sci.* **157,** 758–78.

HOEBEL, B. G. (1971) Feeding: neural control of intake. *Ann. Rev. Physiol.* **33,** 533–68.

HORNYKIEWICZ, O. (1973) Dopamine in the basal ganglia: its role and therapeutic implications (including the use of L-Dopa). *Br. Med. Bull.* **29,** 172–8.

HOUPT, K. A. and EPSTEIN, A. N. (1971) The complete dependence of beta-adrenergic drinking on the renal dipsogen. *Physiol. Behav.* **7,** 897–902.

HOWARTH, C. I. and DEUTSCH, J. A. (1962) Drive decay: the cause of fast "extinction" of habits learned for brain stimulation. *Science* **137,** 35–36.

HUANG, Y. H. and MOGENSON, G. J. (1972) Neural pathways mediating drinking and feeding in rats. *Expl. Neurol.* **37,** 269–86.

HUANG, Y. H. and ROUTTENBERG, A. (1971) Lateral hypothalamic self-stimulation pathways in *Rattus norvegicus*. *Physiol. Behav.* **7,** 419–32.

HUSTON, J. P. and BORBELY, A. A. (1973) Operant conditioning in forebrain ablated rats by use of rewarding hypothalamic stimulation. *Brain Res.* **50,** 467–72.

ITO, M. (1972) Excitability of medial forebrain bundle neurons during self-stimulation behavior. *J. Neurophysiol.* **35,** 652–64.

ITO, M. and OLDS, J. (1971) Unit activity during self-stimulation behavior. *J. Neurophysiol.* **34,** 263–73.

JACKSON, F. (1968) The effects of posterior hippocampal lesions upon intracranial self-stimulation rates under conditions of satiation and deprivation. *Dissert. Abstr.* **29,** 1517B.

JACOBSEN, C. F. (1936) Studies of cerebral functions in primates. *Compar. Psychol. Monogr.* **13,** 1–60.

JANOWITZ, H. D. and GROSSMAN, M. I. (1949) Some factors affecting the food intake of normal dogs and dogs with esophagostomy and gastric fistula. *Am. J. Physiol.* **159,** 143–8.

JANOWITZ, H. D. and HOLLANDER, F. (1953) Effect of prolonged intragastric feeding on oral ingestion. *Fedn. Proc.* **12,** 72.

JONES, B. and MISHKIN, M. (1972) Limbic lesions and the problem of stimulus–reinforcement associations. *Expl. Neurol.* **36,** 362–77.

JONES, B. E., BOBILLIER, P., PIN, C., and JOUVET, M. (1973) The effect of lesions of catecholamine-containing neurons upon monoamine content of the brain and EEG and behavioral waking in the cat. *Brain Res.* **58,** 157–77.

JONES, E. G. and POWELL, T. P. S. (1970) An anatomical study of converging sensory pathways within the cerebral cortex of the monkey. *Brain* **93,** 793–820.

JORDAN, H. A. (1969) Voluntary intragastric feeding: oral and gastric contributions to food intake and hunger in man. *J. Comp. Physiol. Psychol.* **68,** 498–506.

JOUVET, M. (1972) The role of monoamines and acetylcholine containing neurons in the regulation of the sleep–waking cycle. *Ergebn. Physiol.* **64,** 166–307.

JUSTESEN, D. R., SHARP, J. C., and PORTER, P. B. (1963) Self-stimulation of the caudate nucleus by instrumentally naive cats. *J. Comp. Physiol. Psychol.* **56,** 371–4.

KANT, K. J. (1969) Influences of amygdala and medial forebrain bundle on self-stimulation in the septum. *Physiol. Behav.* **4,** 777–84.

KEENE, J. J. and CASEY, K. L. (1973) Rewarding and aversive brain stimulation: opposite effects on medial thalamic units. *Physiol. Behav.* **10,** 283–7.

KEESEY, R. E. (1964) Intracranial reward delay and the acquisition rate of a brightness discrimination. *Science* **143,** 702–3.

KEESEY, R. E. and POWLEY, T. L. (1968) Enhanced lateral hypothalamic reward sensitivity following septal lesions in the rat. *Physiol. Behav.* **3,** 557–62.

KELLICUTT, M. H. and SCHWARTZBAUM, J. S. (1963) Formation of a conditioned emotional response (CER) following lesions of the amygdaloid complex in rats. *Psychol. Rep.* **12,** 351–8.

KELLY, P. H. (1974) The physiological basis of reward and punishment in vertebrates, DPhil. thesis, Oxford University.

KELLY, P. H., ROLLS, E. T., and SHAW, S. G. (1974) Functions of catecholamines in brain-stimulation reward. *Brain Res.* **66,** 363–4.

KENT, R. and GROSSMAN, S. P. (1969) Evidence for a conflict interpretation of anomalous effects of rewarding brain stimulation. *J. Comp. Physiol. Psychol.* **69,** 381–90.

KETY, S. S. (1970) Neurochemical aspects of emotional behavior. In *Physiological Correlates of Emotion*, ch. 4, pp. 61–71 (ed. P. Black), Academic Press, New York.

KLÜVER, H. and BUCY, P. C. (1939) Preliminary analysis of functions of the temporal lobes in monkeys. *Archs Neurol. Psychiat. Chicago* **42,** 979–1000.

LE MAGNEN, J. (1971) Advances in studies on the physiological control and regulation of food intake. In *Progress in Physiological Psychology*, vol. 4, pp. 204–61 (ed. E. Stellar and J. M. Sprague), Academic Press, New York.

LE MAGNEN, J. and VINCENT, J. D. (1973) Electrophysiological correlates of "specific arousals", paper presented at the Workshop Meeting of the European Brain and Behaviour Society, "Behavioural Functions of the Limbic System", Madrid, 24–25 April, 1973.

LEONARD, C. M. (1969) The prefrontal cortex of the rat: I, Cortical projection of the mediodorsal nucleus: II, Efferent connections. *Brain Res.* **12,** 321–43.

LILLY, J. (1960) Learning activated by subcortical stimulation: the "start" and the "stop" patterns of behavior. In *Electrical Studies on the Unanesthetized Brain* (ed. E. R. Ramey and D. S. O'Doherty), Hoeber, New York.

LILLY, J. C. and MILLER, A. M. (1962) Operant conditioning of the bottlenose dolphin with electrical stimulation of the brain. *J. Comp. Physiol. Psychol.* **55,** 73–79.

LIVETT, B. G. (1973) Histochemical visualization of peripheral and central adrenergic neurones. *Br. Med. Bull.* **29,** 93–99.

LORENS, A. S. (1966) Effects of lesions in the central nervous system on lateral hypothalamic self-stimulation in the rat. *J. Comp. Physiol. Psychol.* **62,** 256–62.

LURIA, A. R. (1973) *The Working Brain*, Penguin, Harmondsworth, England.

Machne, X., Calma, I., and Magoun, W. (1955) Unit activity of central cephalic brainstem in EEG arousal. *J. Neurophysiol.* **18,** 547–58.

MacNeil, D. A. (1966) Inhibition of food intake and hypothalamic self-stimulation correlated with excess body weight, PhD thesis, Princeton University, Univ. Microfilm No. 66-13332.

Madlafousek, J., Freund, K., and Grofova, I. (1970) Variables determining the effect of electrostimulation in the lateral preoptic area on the sexual behavior of male rats. *J. Comp. Physiol. Psychol.* **72,** 28–44.

Malmo, R. B. (1961) Slowing of heart rate following septal self-stimulation in rats. *Science* **133,** 1128–30.

Margules, D. L. (1969) Noradrenergic rather than serotonergic basis of reward in dorsal tegmentum. *J. Comp. Physiol. Psychol.* **67,** 32–35.

Margules, D. L. and Olds, J. (1962) Identical "feeding" and "rewarding" systems in the lateral hypothalamus of rats. *Science* **135,** 374–5.

Margules, D. L. and Stein, L. L. (1967) Neuroleptics vs. tranquilizers: evidence from animal behavior studies of the mode and site of action. In *Neuro-psychopharmacology*, pp. 108–20 (ed. H. Brill), Excerpta Medica Foundation, Amsterdam.

Mark, V. H. and Ervin, F. R. (1970) *Violence and the Brain*, Harper & Row, New York.

Mark, V. H., Ervin, F. R., and Sweet, W. H. (1972) Deep temporal lobe stimulation in man. In *The Neurobiology of the Amygdala*, pp. 485–507 (ed. B. E. Eleftheriou), Plenum Press, New York.

Marks, I. M. (1969) *Fears and Phobias*, Heinemann, London.

Marshall, J., Turner, B. H., and Teitelbaum, P. (1971) Sensory neglect produced by lateral hypothalamic damage. *Science* **174,** 523–5.

Melzack, R. (1973) *The Puzzle of Pain*, Penguin, Harmondsworth, England.

Mendelson, J. (1970) Self-induced drinking in rats: qualitative identity of drive and reward systems in the lateral hypothalamus. *Physiol. Behav.* **5,** 925–30.

Meyers, W. J., Valenstein, E. S., and Lacey, J. I. (1963) Heart rate changes after reinforcing brain stimulation in rats. *Science* **140,** 1233–5.

Milgram, N. W. (1969) Effect of hippocampal stimulation on feeding in the rat. *Physiol. Behav.* **4,** 665–70.

Milner, P. M. (1970) *Physiological Psychology*, Holt, Rinehart, & Winston, New York.

Miselis, R. and Epstein, A. N. (1971) Preoptic-hypothalamic mediation of feeding induced by cerebral glucoprivation. *Am. Zool.* **11,** 624.

Mishkin, M. (1966) Visual mechanisms beyond the striate cortex. In *Frontiers in Physiological Psychology*, vol. 4, pp. 93–119 (ed. R. W. Russell), Academic Press, New York.

Mishkin, M. (1970) Cortical visual areas and their interactions. In *The Brain and Human Behavior* (ed. A. G. Karczmar), Springer, Berlin.

Mogenson, G. J. (1964) Effects of sodium pentobarbital on brain self-stimulation. *J. Comp. Physiol. Psychol.* **58,** 461–2.

Mogenson, G. J. (1965) An attempt to establish secondary reinforcement with rewarding brain self-stimulation. *Psychol. rep.* **16,** 163–7.

Mogenson, G. J. (1969) Water deprivation and excessive water intake during self-stimulation. *Physiol. Behav.* **4,** 393–7.

Moniz, E. (1936) *Tentatives opératoires dans le traitement de certaines psychoses*, Masson, Paris.

MOORE, K. E. (1971) Biochemical correlates of the behavioral effects of drugs, ch. 3 in *An Introduction to Psychopharmacology* (eds. R. H. Rech and K. E. Moore), Raven Press, New York.

MOUNT, G. and HOEBEL, B. G. (1967) Lateral hypothalamic self-stimulation: self-determined threshold increased by food intake. *Psychon. Sci.* **9,** 265–6.

MUSACCHIO, J. M., GOLDSTEIN, M., ANAGNOSTE, B., POCH, G., and KOPIN, I. J. (1966) Inhibition of dopamine-β-hydroxylase by disulfiram *in vivo*. *J. Pharmac. Exp. Ther.* **152,** 56–66.

NARABAYASHI, H. (1972) Stereotaxic amygdalotomy. In *The Neurobiology of the Amygdala*, pp. 459–83 (ed. B. E. Eleftheriou), Plenum Press, New York.

NAUTA, W. J. H. (1971) The problem of the frontal lobe: a reinterpretation. *J. Psychiat. Res.* **8,** 167–87.

NEWMAN, L. M. (1972) Effects of cholinergic agonists and antagonists on self-stimulation behavior in the rat. *J. Comp. Physiol. Psychol.* **79,** 394–413.

NICKERSON, M. and HOLLENBERG, N. K. (1967) Blockade of α-adrenergic receptors. In *Physiological Pharmacology*, vol. 4, pp. 243–305 (ed. W. S. Root and F. G. Hoffman), Academic Press, New York.

NICOLAIDIS, S. (1969) Early systemic responses to orogastric stimulation in the regulation of food and water balance: functional and electrophysiological data. *Ann. NY Acad. Sci.* **157,** 1176–1203.

O'KEEFE, J. and BOUMA, H. (1969) Complex sensory properties of certain amygdala units in the freely moving cat. *Expl. Neurol.* **23,** 384–98.

OLDS, J. (1956) Runway and maze behavior controlled by basomedial forebrain stimulation in the rat. *J. Comp. Physiol. Psychol.* **49,** 507–12.

OLDS, J. (1958) Effects of hunger and male sex hormone on self-stimulation of the brain. *J. Comp. Physiol. Psychol.* **51,** 320–4.

OLDS, J. (1960) Differentiation of reward systems in the brain by self-stimulation techniques. In *Electrical Studies on the Unanesthetized Brain* (ed. E. R. Ramey and D. S. O'Doherty), Hoeber, New York.

OLDS, J. (1961) Differential effects of drive and drugs on self-stimulation at different brain sites. In *Electrical Stimulation of the Brain* (ed. D. E. Sheer), University of Texas Press, Austin.

OLDS, J. (1962) Hypothalamic substrates of reward. *Physiol. Rev.* **42,** 554–604.

OLDS, J. and MILNER, P. (1954) Positive reinforcement produced by electrical stimulation of septal area and other regions of the rat brain. *J. Comp. Physiol. Psychol.* **47,** 419–27.

OLDS, J. and OLDS, M. (1965) Drives, rewards, and the brain. In *New Directions in Psychology*, vol. II, pp. 327–410 (ed. F. Barron and W. C. Dement), Holt, Rinehart, & Winston, New York.

OLDS, J., TRAVIS, R. P., and SCHWING, R. C. (1960) Topographic organisation of hypothalamic self-stimulation functions. *J. Comp. Physiol. Psychol.* **53,** 23–32.

OLDS, J., ALLAN, W. S., and BRIESE, A. E. (1971) Differentiation of hypothalamic drive and reward centers. *Am. J. Physiol.* **221,** 368–75.

OLDS, M. E. (1961) Effects of lesions in medial forebrain bundle on self-stimulation behavior. *Am. J. Physiol.* **217,** 1253–64.

OLDS, M. E. (1970) Comparative effects of amphetamine, scopolamine, chlordiazepoxide, and diphenylhydantoin on operant and extinction behavior with brain stimulation and food reward. *Neuropharmacology* **9,** 519–32.

OLDS, M. E. (1973) Short-term changes in the firing pattern of hypothalamic neurons during Pavlovian conditioning. *Brain Res.* **58,** 95–116.

OLDS, M. E. and OLDS, J. (1962) Approach–escape interactions in rat brain. *Am. J. Physiol.* **203,** 803–10.

OLDS, M. E. and OLDS, J. (1969) Effects of lesions in medial forebrain bundle on self-stimulation behavior. *Am. J. Physiol.* **217,** 1253–64.

PANKSEPP, J. and TROWILL, J. A. (1967a) Intraoral self-injection: I, Effects of delay of reinforcement on resistance to extinction and implications for self-stimulation. *Psychon. Sci.* **9,** 405–6.

PANKSEPP, J. and TROWILL, J. A. (1967b) Intraoral self-injection: II, The simulation of self-stimulation phenomena with a conventional reward. *Psychon. Sci.* **9,** 407–8.

PENFIELD, W. and JASPER, H. (1954) *Epilepsy and the Functional Anatomy of the Human Brain,* Little, Brown, Boston.

PEREZ-CRUET, J., BLACK, W. C., and BRADY, J. V. (1963) Heart rate: differential effects of hypothalamic and septal self-stimulation. *Science* **140,** 1235–6.

PEREZ-CRUET, J., MCINTIRE, R. W., and PLISKOFF, S. S. (1965) Blood pressure and heart-rate changes in dogs during hypothalamic self-stimulation. *J. Comp. Physiol. Psychol.* **60,** 373–81.

PHILLIPS, A. G. (1970) Enhancement and inhibition of olfactory bulb self-stimulation by odours. *Physiol. Behav.* **5,** 1127–31.

PHILLIPS, A. G. and FIBIGER, H. C. (1973) Dopaminergic and noradrenergic substrate of positive reinforcement. *Science* **179,** 575–6.

PHILLIPS, A. G. and MOGENSON, G. J. (1969) Self-stimulation of the olfactory bulb. *Physiol. Behav.* **4,** 195–7.

PORTER, R. W., CONRAD, D. G., and BRADY, J. V. (1959) Some neural and behavioral correlates of electrical self-stimulation of the limbic system. *J. Exp. Anal. Behav.* **2,** 43–55.

POSCHEL, B. P. H. and NINTEMAN, F. W. (1971) Intracranial reward and the forebrain's serotonergic mechanism: studies employing *para*-chlorophenylalanine and *para*-chloroamphetamine. *Physiol. Behav.* **7,** 39–46.

POWELL, T. P. S., COWAN, W. M., and RAISMAN, G. (1965) The central olfactory connexions. *J. Anat.* **99,** 791–813.

QUARTERMAIN, D. and WEBSTER, D. (1968) Extinction following intracranial reward: the effect of delay between acquisition and extinction. *Science* **159,** 1259–60.

RAISMAN, G. (1966) Neural connexions of the hypothalamus. *Br. Med. Bull.* **22,** 197–201.

RECH, R. H. and MOORE, K. E. (eds.) (1971) *An Introduction to Psychopharmacology.* Raven Press, New York.

REID, L. D., GIBSON, W. E., GLEDHILL, S. M., and PORTER, P. B. (1964) Anticonvulsant drugs and self-stimulation behavior. *J. Comp. Physiol. Psychol.* **57,** 353–6.

REID, L. D., HUNSICKER, J. P., KENT, E. W., LINDSAY, J. I., and GALLISTEL, C. E. (1973) Incidence and magnitude of the "priming effect" in self-stimulating rats. *J. Comp. Physiol. Psychol.* **82,** 286–93.

RITTER, S. and STEIN, L. (1972) Self-stimulation of the locus coeruleus. *Fedn. Proc.* **31,** 820.

ROBERTS, W. W. (1958) Both rewarding and punishing effects from stimulation of posterior hypothalamus of cat with same electrode at same intensity. *J. Comp. Physiol. Psychol.* **51,** 400–7.

ROLL, S. K. (1970) Intracranial self-stimulation and wakefulness: effects of manipulating ambient brain catecholamines. *Science* **168,** 1370–2.

ROLLS, B. J. and ROLLS, E. T. (1973) Effects of lesions in the basolateral amygdala on fluid intake in the rat. *J. Comp. Physiol. Psychol.* **83,** 240–7.

Rolls, E. T. (1970a) Neural systems involved in intracranial self-stimulation. *Brain Res.* **24,** 548.

Rolls, E. T. (1970b) Neural mechanisms of intracranial self-stimulation in the rat, DPhil. thesis, Oxford University.

Rolls, E. T. (1971a) Involvement of brainstem units in medial forebrain bundle self-stimulation. *Physiol. Behav.* **7,** 297–310.

Rolls, E. T. (1971b) Absolute refractory period of neurons involved in MFB self-stimulation. *Physiol. Behav.* **7,** 311–15.

Rolls, E. T. (1971c) Contrasting effects of hypothalamic and nucleus accumbens septi self-stimulation on brain stem single unit activity and cortical arousal. *Brain Res.* **31,** 275–85.

Rolls, E. T. (1971d) Absolute refractory period of neurones involved in drinking elicited by electrical stimulation of the lateral hypothalamus. *J. Physiol. Lond.*, **218,** 46–47P.

Rolls, E. T. (1972) Activation of amygdaloid neurones in reward, eating and drinking elicited by electrical stimulation of the brain. *Brain Res.* **45,** 365–81.

Rolls, E. T. (1973) Refractory periods of neurons involved in stimulus-bound eating and drinking in the rat. *J. Comp. Physiol. Psychol.* **82,** 15–22.

Rolls, E. T. (1974) The neural basis of brain-stimulation reward. *Prog. Neurobiol.* **3,** 71–160.

Rolls, E. T. and Cooper, S. J. (1973) Activation of neurones in the prefrontal cortex by brain-stimulation reward in the rat. *Brain Res.* **60,** 351–68.

Rolls, E. T. and Cooper, S. J. (1974a) Anesthetization and stimulation of the sulcal prefrontal cortex and brain-stimulation reward. *Physiol. Behav.* **12,** 563–71.

Rolls, E. T. and Cooper, S. J. (1974b) Connection between prefrontal cortex and pontine brain-stimulation reward sites in the rat. *Expl. Neurol.* **42,** 687–99.

Rolls, E. T. and Kelly, P. H. (1972) Neural basis of stimulus-bound locomotor activity in the rat. *J. Comp. Physiol. Psychol.* **81,** 173–82.

Rolls, E. T. and Rolls, B. J. (1973) Altered food preferences after lesions in the basolateral region of the amygdala in the rat. *J. Comp. Physiol. Psychol.* **83,** 248–59.

Rolls, E. T., Kelly, P. H., and Shaw, S. G. (1974a) Noradrenaline, dopamine and brain-stimulation reward. *Pharmac. Biochem. Behav.* (in press).

Rolls, E. T., Rolls, B. J., Kelly, P. H., Shaw, S. G., and Dale, R. (1974b) The relative attenuation of self-stimulation, eating and drinking produced by dopamine-receptor blockade. *Psychopharmacologia* **38,** 219–30.

Routtenberg, A. (1970) Hippocampal activity and brainstem reward–aversion loci. *J. Comp. Physiol. Psychol.* **72,** 161–70.

Routtenberg, A. (1971) Forebrain pathways of reward in *Rattus norvegicus. J. Comp. Physiol. Psychol.* **75,** 269–76.

Routtenberg, A. and Huang, Y. H. (1968) Reticular formation and brainstem unitary activity: effects of posterior hypothalamic and septal-limbic stimulation at reward loci. *Physiol. Behav.* **3,** 611–17.

Routtenberg, A. and Malsbury, C. (1969) Brainstem pathways of reward. *J. Comp. Physiol. Psychol.* **68,** 22–30.

Routtenberg, A. and Olds, J. (1963) The attenuation of response to an aversive brain stimulus by concurrent rewarding septal stimulation. *Fedn. Proc.* **22,** 215.

Routtenberg, A. and Sloan, M. (1972) Self-stimulation in the frontal cortex of *Rattus norvegicus. Behav. Biol.* **7,** 567–72.

Routtenberg, A., Gardner, E. L., and Huang, Y. H. (1971) Self-stimulation pathways in the monkey, *Macaca mulatta. Expl. Neurol.* **33,** 213–24.

ROWLAND, N. E. (1973) Systemic factors in the control of water intake in the rat, PhD thesis, University of London.

RYLANDER, G. (1948) Personality analysis before and after frontal lobotomy. *Ass Res. Nerv. Ment. Dis., The Frontal Lobes* **27,** 691–705.

SACKS, O. (1973) *Awakenings*, Duckworth, London.

SCHIFF, B. W. (1964) The effects of tegmental lesions on the reward properties of septal stimulation.

SCHILDKRAUT, J. J. (1965) The catecholamine hypothesis of affective disorders: review of supporting evidence. *Am. J. Psychiat.* **122,** 509–22.

SCHILDKRAUT, J. J. and KETY, S. S. (1967) Biogenic amines and emotion. *Science* **156,** 21–30.

SEM-JACOBSEN, C. W. (1968) *Depth-electrographic Stimulation of the Human Brain and Behavior: From Fourteen Years of Studies and Treatment of Parkinson's Disease and Mental Disorders with Implanted Electrodes*, C. C. Thomas, Springfield, Ill.

SEM-JACOBSEN, C. W. and TORKILDSEN, A. (1960) Depth recording and electrical stimulation in the human brain. In *Electrical Studies on the Unanesthetized Brain*, pp. 275–310 (ed. E. R. Ramey and D. S. O'Doherty), Hoeber, New York.

SEWARD, J. P., UYEDA, A. A., and OLDS, J. (1959) Resistance to extinction following cranial self-stimulation. *J. Comp. Physiol. Psychol.* **52,** 294–9.

SHUTE, C. C. D. and LEWIS, P. R. (1966) Cholinergic and monoamine pathways in the hypothalamus. *Br. Med. Bull.* **22,** 221–6.

SPIEGEL, T. A. (1973) Caloric regulation of food intake in man. *J. Comp. Physiol. Psychol.* **84,** 24–37.

STARK, P. and BOYD, E. S. (1961) Electrical self-stimulation by dogs through chronically implanted electrodes in the hypothalamus. *Fedn. Proc.* **20,** 328.

STARK, P. and FULLER, R. W. (1972) Behavioral and biochemical effects of *p*-chlorophenylalanine, 3-chlorotyrosine and 3-chlorotyramine: a proposed mechanism for inhibition of self-stimulation. *Neuropharmacology* **11,** 261–72.

STEFFENS, A. B. (1970) Plasma insulin content in relation to blood glucose level and meal pattern in the normal and hypothalamic hyperphagic rat. *Physiol. Behav.* **5,** 147–51.

STEIN, L. (1965) Facilitation of avoidance behavior by positive brain stimulation. *J. Comp. Physiol. Psychol.* **60,** 9–19.

STEIN, L. (1967) Psychopharmacological substrates of mental depression. In *Antidepressant Drugs* (ed. S. Garattini and M. N. G. Dukes), Excerpta Medica Foundation, Amsterdam.

STEIN, L. (1969) Chemistry of purposive behavior. In *Reinforcement and Behavior*, pp. 328–35 (ed. J. Tapp), Academic Press, New York.

STEIN, L. (1971) Neurochemistry of reward and punishment: some implications for the etiology of schizophrenia. *J. Psychiat. Res.* **8,** 345–61.

STEIN, L. and RAY, O. S. (1959) Self-regulation of brain stimulating current intensity in the rat. *Science* **130,** 570–2.

STEIN, L. and WISE, C. D. (1969) Release of norepinephrine from hypothalamus and amygdala by rewarding stimulation of the medial forebrain bundle. *J. Comp. Physiol. Psychol.* **67,** 189–98.

STEIN, L. and WISE, C. D. (1971) Possible etiology of schizophrenia: progressive damage to the noradrenergic reward system by 6-hydroxydopamine. *Science* **171,** 1032–6.

TANAKA, D. (1973) Effects of selective prefrontal decortication on escape behavior in the monkey. *Brain Res.* **53,** 161–73.

TENEN, S. S. (1967) The effects of *p*-chlorophenylalanine, a serotonin depletor on avoidance acquisition, pain sensitivity and related behavior in the rat. *Psychopharmacologia* **10,** 204–19.

TEUBER, H. L. (1972) Unity and diversity of frontal lobe functions. In *The Frontal Granular Cortex and Behavior* (International Symposium), *Acta Neurobiologiae Experimentalis Warsaw* **32,** 615–56 (ed. J. Konorski, H. L. Teuber, and B. Zernicki).

THOMPSON, R. F. (1967) *Foundations of Physiological Psychology,* Harper & Row, New York.

UNGERSTEDT, U. (1971) Stereotaxic mapping of the monoamine pathways in the rat brain. *Acta physiol. scand.* Suppl. **367,** 1–48.

URSIN, R., URSIN, H., and OLDS, J. (1966) Self-stimulation of hippocampus in rats. *J. Comp. Physiol. Psychol.* **61,** 353–9.

VALENSTEIN, E. S. (1966) The anatomical locus of reinforcement. *Prog. Physiol. Psychol.* **1,** 149–90.

VALENSTEIN, E. S. (1974) *Brain Control. A Critical Examination of Brain Stimulation and Psychosurgery,* Wiley, New York.

VALENSTEIN, E. S. and CAMPBELL, J. F. (1966) Medial forebrain bundle—lateral hypothalamic area and reinforcing brain stimulation. *Am. J. Physiol.* **210,** 270–4.

VALENSTEIN, E. S. and VALENSTEIN, T. (1964) Interaction of positive and negative reinforcing neural systems. *Science* **145,** 1456–7.

VALENSTEIN, E. S., COX, V. C., and KAKOLEWSKI, J. J. (1969) The hypothalamus and motivated behavior. In *Reinforcement and behavior* (ed. J. Tapp), Academic Press, New York.

VALENSTEIN, E. S., COX, V. C., and KAKOLEWSKI, J. W. (1970) Re-examination of the role of the hypothalamus in motivation. *Psychol. Rev.* **77,** 16–31.

WALIKE, B. C., JORDAN, H. A., and STELLAR, E. (1969) Preloading and the regulation of food intake in man. *J. Comp. Physiol. Psychol.* **68,** 327–33.

WARD, H. P. (1960) Basal tegmental self-stimulation after septal ablation in rats. *AMA Archs. Neurol.* **3,** 158–62.

WARD, H. P. (1961) Tegmental self-stimulation after amygdaloid ablation. *AMA Archs. Neurol.* **4,** 657–9.

WARD, J. W. and HESTER, R. W. (1969) Intracranial self-stimulation in cats surgically deprived of autonomic outflows. *J. Comp. Physiol. Psychol.* **67,** 336–43.

WAUQUIER, A. and NIEMEGEERS, C. J. E. (1972) Intracranial self-stimulation in rats as a function of various stimulus parameters: II, Influence of Haloperidol, Pimozide and Pipamperone on medial forebrain stimulation with monopolar electrodes. *Psychopharmacologia* **27,** 191–202.

WEISKRANTZ, L. (1956) Behavioral changes associated with ablation of the amygdaloid complex in monkeys. *J. Comp. Physiol. Psychol.* **49,** 381–91.

WEISKRANTZ, L. (1972) Behavioural analysis of the monkey's visual nervous system. *Proc. R. Soc. Lond.* B, **182,** 427–55.

WETZEL, M. C., HOWELL, L. G., and BEARIE, K. J. (1969) Experimental performance of steel and platinum electrodes with chronic monophasic stimulation of the brain. *J. Neurosurg.* **31,** 658–69.

WISE, C. D. and STEIN, L. (1969) Facilitation of brain self-stimulation by central administration of norepinephrine. *Science* **163,** 299–301.

WOOLEY, O. W., WOOLEY, S. C., and DUNHAM, R. B. (1972) Calories and sweet taste: effects on sucrose preference in the obese and nonobese. *Physiol. Behav.* **9,** 765–8.

WURTZ, R. H. and OLDS, J. (1963) Amygdaloid stimulation and operant reinforcement in the rat. *J. Comp. Physiol. Psychol.* **56,** 941–9.

ZEMAN, W. and INNES, J. R. M. (1963) *Craigie's Neuroanatomy of the Rat,* Academic Press, New York.

INDEX